Recent Advances in Heat Pipes

Edited by Wael I. A. Aly

Published in London, United Kingdom

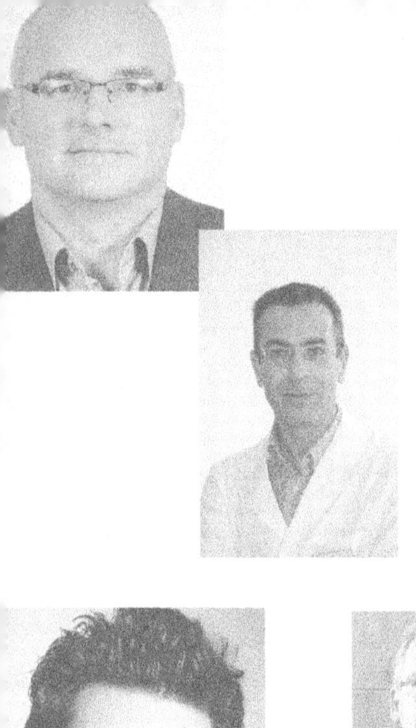

IntechOpen

Supporting open minds since 2005

Recent Advances in Heat Pipes
http://dx.doi.org/10.5772/intechopen.75280
Edited by Wael I. A. Aly

Contributors
Chunsheng Guo, Bala Abdullahi, Raya K. Al-dadah, Sa'ad Mahmoud, Xudong Zhao, Xingxing Zhang, Zhangyuan Wang, Xinru Wang, Chuangbin Weng, Wael I.A. Aly

First published in London, United Kingdom, 2019 by IntechOpen
IntechOpen is the global imprint of INTECHOPEN LIMITED, registered in England and Wales, registration number: 11086078, The Shard, 25th floor, 32 London Bridge Street
London, SE19SG - United Kingdom
Printed in Croatia

British Library Cataloguing-in-Publication Data
A catalogue record for this book is available from the British Library

Additional hard and PDF copies can be obtained from orders@intechopen.com

Recent Advances in Heat Pipes
Edited by Wael I. A. Aly
p. cm.
Print ISBN 978-1-83962-643-2
Online ISBN 978-1-83962-208-3
eBook (PDF) ISBN 978-1-83962-209-0

Meet the editor

Wael I. A. Aly is Professor of Mechanical Engineering and Head of the Department of Refrigeration and Air Conditioning Technology, Faculty of Industrial Education, Helwan University. Dr. Wael obtained his PhD (2007) from Okayama University, Japan, in the field of Thermofluids. His MSc (1997) was in the field of RHVAC from TUE, The Netherlands. He obtained his BSc (1994) in Mechanical Engineering from Benha University, Egypt. He is a certified reviewer for 20 international journals. Dr. Wael's current research work is in the fields of applications of nanofluids in thermal systems, heat transfer enhancement and heat pipes, absorption and adsorption refrigeration using solar power and engine exhaust gases, turbulent flow drag reduction, and computational fluid dynamics. He has published more than 35 papers.

Contents

Preface XI

Chapter 1 1
Introductory Chapter: Recent Advances in Heat Pipes
by Wael I.A. Aly

Chapter 2 5
Thermosyphon Heat Pipe Technology
by Bala Abdullahi, Raya K. Al-dadah and Sa'ad Mahmoud

Chapter 3 25
The Recent Research of Loop Heat Pipe
by Chunsheng Guo

Chapter 4 49
Study of a Novel Liquid-Vapour Separator-Incorporated Gravitational Loop Heat Pipe
by Xudong Zhao, Chuangbin Weng, Xingxing Zhang, Zhangyuan Wang and Xinru Wang

Preface

Throughout the years, new applications for heat pipes have been discovered. Heat pipes are considered as an effective thermal solution, particularly in high heat flux applications and in situations where there is a combination of nonuniform heat loading, limited airflow over the heat-generating components, and space or weight constraints.

This book is intended to explore some of the recent developments and advances in heat pipes and their applications in thermal systems. The book starts with experimental and numerical thermosyphon heat pipe technology, and then looks at heat pipes and loop heat pipes.

The book should be of interest to engineers and those who intend to do research in this exciting field.

Wael I.A. Aly
Professor of Mechanical Engineering,
Department of Refrigeration and Air Conditioning Technology,
Faculty of Industrial Education,
Helwan University,
Egypt

Chapter 1

Introductory Chapter: Recent Advances in Heat Pipes

Wael I.A. Aly

1. Introduction

Heat pipe is a two-phase flow passive and reliable heat transfer device widely used in thermal systems [1]. It is known that the thermal conductance of heat pipes is higher than any solid conductor due to the accompanying latent heat during the closed two-phase cycle. Moreover, heat pipes have many advantages compared to other heat exchangers: higher amounts of heat transferred over long distance, faster thermal response time, easier design and manufacturing, lower temperature difference, broad temperature range for applications, and easier control which allow transporting high rates of heat at various temperature levels. Also, as a passive device, no external power is required for its operation, and heat pipe is highly reliable and almost requires no maintenance. Because of the mentioned advantages, heat pipes are ideal for many applications. Heat pipe is considered as an effective thermal solution, particularly in high heat flux applications and in situations where there is a combination of nonuniform heat loading, limited airflow over the heat-generating components, and space or weight constraints.

After the introduction of heat pipes with the paper *Structures of Very High Thermal Conductance* by the authors Grover et al. [2] in 1964, the interest in the applications of heat pipes has increased remarkably. Currently, a huge amount of documents (research articles, review articles, and books) concerning heat pipes

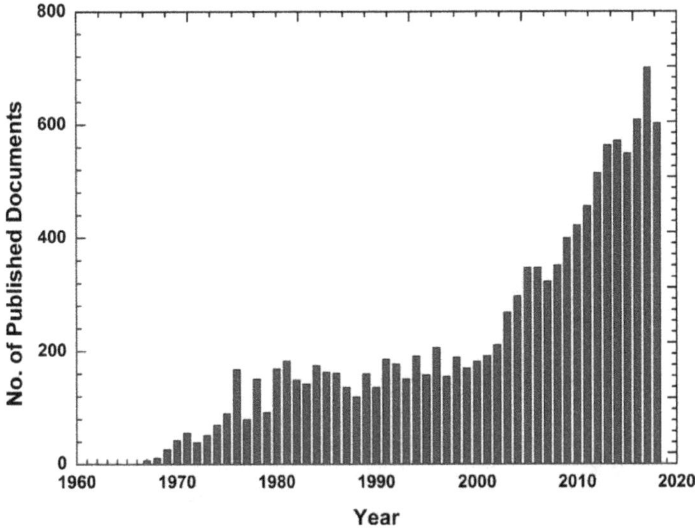

Figure 1.
Number of published documents on heat pipes each year.

and two-phase closed thermosyphon are published. The literature is very rich now with documents about heat pipes, and the heat pipe has become recognized as an important development in heat transfer technology. Many researchers assessed the potential applications of various types of heat pipes. **Figure 1** shows how the research on heat pipes evolved from its introduction in 1964 up to the present time. Around 12,372 documents have been published since 1964. The data were extracted from the Scopus database by searching “heat pipe” or “heat pipes” in the article title, abstract, and keywords (date of extract: 25 May 2019). As shown in **Figure 1**, after the year 2000, the number of documents per year increased remarkably, so that by 2017, it reached more than 700 papers.

2. Advances on heat pipes

The recent advances of heat pipes may include recent advances in working fluids (nanofluids, new refrigerants, etc.), wick structures (microgrooves, sintered, etc.), special types of heat pipes (VCHP, pulsating HP, rotating HP, electrokinetic force), and new applications (energy conservation and storage, reactors, spacecraft, renewable energy, food industries, cooling of electronic components, etc.) [3–5].

Author details

Wael I.A. Aly
Department of Refrigeration and Air Conditioning Technology, Faculty of Industrial Education, Helwan University, Egypt

*Address all correspondence to: aly_wael@helwan.edu.eg; aly_wael@yahoo.com

References

[1] Aly WI, Elbalshouny MA, El-Hameed HA, Fatouh M. Thermal performance evaluation of a helically-micro-grooved heat pipe working with water and aqueous Al_2O_3 nanofluid at different inclination angle and filling ratio. Applied Thermal Engineering. 2017;**110**:1294-1304

[2] Grover G, Cotter T, Erickson G. Structures of very high thermal conductance. Journal of Applied Physics. 1964;**35**(6):1990-1991

[3] Alhuyi Nazari M, Ahmadi MH, Ghasempour R, Shafii MB. How to improve the thermal performance of pulsating heat pipes: A review on working fluid. Renewable and Sustainable Energy Reviews. 2018;**91**:630-638

[4] Su Q, Chang S, Zhao Y, Zheng H, Dang C. A review of loop heat pipes for aircraft anti-icing applications. Applied Thermal Engineering. 2018;**130**:528-540

[5] Poplaski LM, Benn SP, Faghri A. Thermal performance of heat pipes using nanofluids. International Journal of Heat and Mass Transfer. 2017;**107**:358-371

Chapter 2

Thermosyphon Heat Pipe Technology

Bala Abdullahi, Raya K. Al-dadah and Sa'ad Mahmoud

Abstract

Heat pipes play vital roles in increasing heat transfer performance of many engineering systems such as solar collectors and this leads to an increase in their usage. Investigation on the performance of heat pipes under different operation conditions and inclination angles is required for effective utilization. In this chapter, a general overview on the construction, operation, advantages, and classifications of heat pipes is presented. Particular attention is given to the heat pipe without wick material in the inner diameter (thermosyphon). Intensive discussions are presented on the construction, operations, advantages and applications of thermosyphon heat pipe. The experimental and numerical approaches on the performance evaluation and characterization of thermosyphon are discussed. A detailed procedure on how experimental work is carried out on thermosyphon is discussed including instrumentation and calibration of the devices. Modelling and simulation of the performance of thermosyphon are discussed, including the model set-up procedure. Factors affecting the performance of thermosyphon such as fill ratio, working fluid, heat input, inclination angles, are analysed based on the overall thermal resistance and thermosyphon performance. Current researches on the effects of major factors affecting the operation of thermosyphon are presented, as well as their current development and various applications in engineering systems.

Keywords: thermosyphon, evaporator, condenser, thermal resistance, inclination angle

1. Introduction

The world's needs for effective heat transfer devices/mechanisms are increasing so as to minimize heat losses, minimize systems cost, enhance heat removal and transportation as well as to increase lifespan of some devices. In some instances, heat is required to be removed from a system (like solar photovoltaic, electrical devices, turbine blades, etc.) in order to keep it at a certain operation temperature, while in other cases, it is required to be transferred to a certain region to keep it at high temperature. Some elements/metals such as copper and aluminium are found to be good conductors of heat as they transfer heat effectively from one region to another. Their ability to transfer heat effectively is due to their molecular arrangements and type of bonds between their molecules. Various systems such as aircraft, electronics, heat exchangers, solar collectors, etc. require effective means of heat transfer. One of the devices recognized as effective means of heat transfer is heat pipe, whose idea was introduced by Graugler in 1942, but its first unit was invented by Grover in 1962;

then, its important properties were studied and identified, and its development started [1]. Hence, with the growing need for efficient heat transfer devices, interest in the use of heat pipes for various applications is increasing due to the roles they play in improving the thermal performance of solar collectors and heat exchangers particularly in energy savings and increasing efficiency of the systems.

Heat pipe is an efficient two-phase heat transfer device which uses latent heat of fluids to transfer energy from one place to another by means of simultaneous evaporation and condensation in a sealed container. It consists of evaporator and condenser sections with or without adiabatic section in between them. Depending on the type, heat pipe may have wick materials on its internal surface where the simultaneous evaporation and condensation take place in the wick structure. In such types of heat pipe, evaporator section can be placed at the top, since the wick structure can return the condensate from the condenser section against gravity. Hence, in a wick heat pipe, the condensed liquid is returned to the evaporator by capillary effects with the assistance of the wick materials as shown in **Figure 1**.

However, many applications do not require inserting wick material on the inner surface of the pipe, because the condenser section can be placed at the top, so that the condensed liquid returns to the evaporator by gravity. This type of wickless heat pipe is called thermosyphon as shown in **Figure 2** Hence, for thermosyphon, the condenser must be above the evaporator, while for the wick heat pipe, the capillary forces in the wick ensure the condensate returns to the evaporator regardless of its position.

1.1 Working principles of heat pipe

Heat pipes consist of sealed vessel usually made from aluminium or copper with or without wick material lined on the inner surface and working fluid charged under a vacuum condition. It is made up of two main sections: evaporator, where the working fluid absorbs heat, and condenser, where the working fluid rejects heat (**Figures 1** and **2**). As heat is added to the working fluid in the evaporator section, it evaporates into vapour when it reaches its saturation temperature. It rises to the condenser with the assistance of buoyancy force and due to the vapour pressure difference between the two sections. The liquid condenses by giving out its

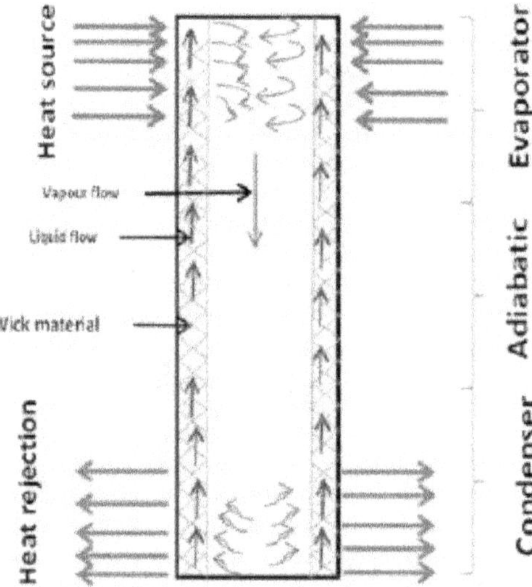

Figure 1.
Operation of wick heat pipe [2].

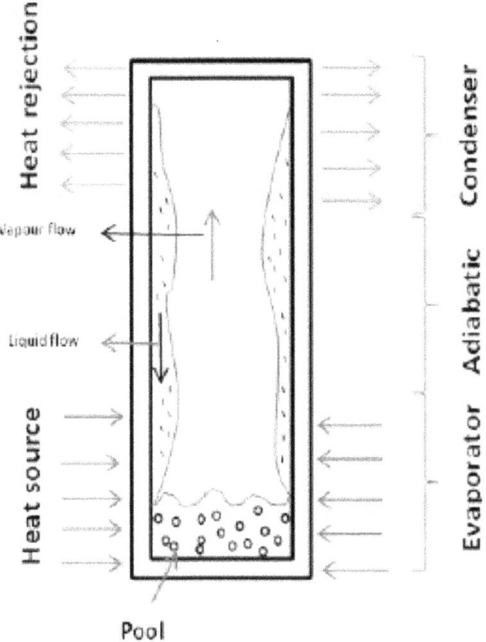

Figure 2.
Operation of thermosyphon [2].

enthalpy to the cooling water in the condenser section and returns back to the evaporator for another cycle.

1.2 Advantages of heat pipe

Heat pipes offer advantages over other heat transfer devices used for various applications in engineering systems. The technology has undergone rapid development due to their operational advantages [3]. Some of these advantages include:

i. High thermal conductivity: In terms of heat transfer, heat pipes are better than the best conductor; hence, they are referred to as 'superconductors'.

ii. Light weight.

iii. Efficient heat transfer.

iv. Flexibility in design.

v. Isothermal operation.

vi. Tolerance to freezing, shock and vibration.

vii. Low cost.

1.3 Classifications of heat pipe

There are different types of heat pipes, classified based on [4]:

I. Nature of fluid circulation, such as capillary driven, rotating heat pipes, flat plate, two-phase close thermosyphon, etc.

II. Control of heat transfer: They are 'controlled heat pipes', such as variable-conductive, thermal switch and thermal diode.

III. Electrostatics-driven heat pipes such as electro hydrodynamic heat pipe.

IV. Osmosis-driven heat pipe such as osmotic heat pipe.

V. Others including inverse, micro, reciprocating, cryogenic, capillary pumped loop heat pipes, etc.

1.4 Applications of heat pipe

Due to the advantages of heat pipes, the technology found its applications in many fields of engineering such as:

i. Spacecraft thermal control [5]: the first test of heat pipe in space was in 1967 [6] and the first heat pipe used for satellite thermal control was on GEOS-B launched from Vanderburgh Air force Base in 1968 [7].

ii. Component cooling, temperature control and radiator design in satellites. Other applications include moderator cooling, removal of heat from the reactor at emitter temperature and elimination of troublesome thermal gradients along the emitter and collector in spacecraft.

iii. Heat pipes for dehumidification and air conditioning: The heat pipe is designed to have one section in the warm incoming stream and the other in the cold outgoing stream. By transferring heat from the warm return air to the cold supply air, the heat pipes create the double effect of pre-cooling the air before it goes to the evaporator and then re-heating it immediately.

iv. Heat exchangers [8].

v. Solar energy systems [9, 10] as shown in **Figure 3**.

vi. Electronic cooling [12, 13], etc.

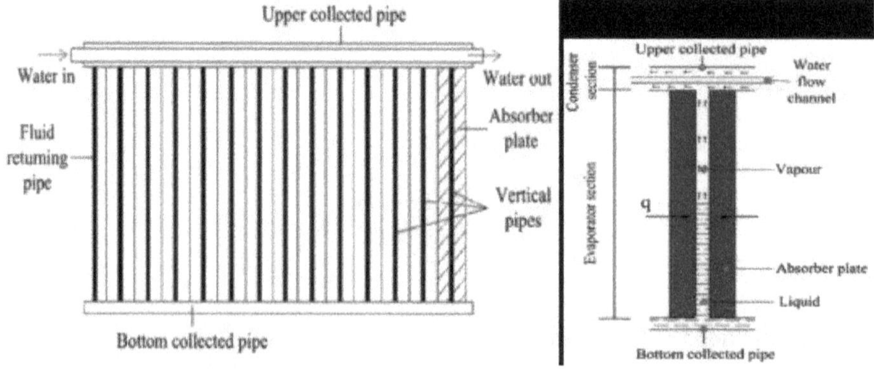

Figure 3.
Developed thermosyphon heat pipe solar collector [11].

1.5 Difference between wick heat pipe and thermosyphon

The wick and wickless (thermosyphon) heat pipes have many features in common in their construction, operation and applications. However, they differ in some aspects such as:

a. Wick material: unlike in thermosyphon, wick materials are lined on the inner parts of the wick heat pipe. This enables the return of the condensed liquid even against gravity.

b. Orientation of the pipes: the condenser section of the thermosyphon must be located at the top of the evaporator because the return of the condensate is basically by gravity, while in the case of the wick heat pipe, the evaporator can be placed at the top because the return of the condensate is based on the capillary effects due to the presence of the wick materials.

c. Need of adiabatic section: thermosyphon may or may not have adiabatic section whereas most of the wick heat pipes have it, as to separate the evaporation and condensing sections.

d. When working fluid is charged into the sealed container, it forms a liquid pool (in case of thermosyphon) while in case of wick heat pipe, it saturates the wick materials.

2. Thermosyphon heat pipe

This is a natural fluid circulation heat pipe which has no wick material presence. It is a simple heat pipe consisting of a sealed vessel charged with working fluid under a vacuum condition. It is made up of evaporator and condenser sections, sometimes with adiabatic section in between them. The vessel is usually made from aluminium or copper to facilitate high conduction of heat. Unlike wick heat pipe, the condenser of thermosyphon must be at the top, for the condensed liquid to return to the evaporator under gravity. Furthermore, some applications of thermosyphon require that the pipe be tilted to an angle from the horizontal for it to have maximum exposure to solar radiation [9, 14–16].

2.1 Construction of thermosyphon heat pipe

Thermosyphon is a vessel closed at both ends and attached with a small charging pipe placed at one of the ends. The air in the vessel is evacuated creating a vacuum, then charged with working fluid through the charging pipe. The pipe is usually divided into the following sections:

i. Evaporator, where heat is supplied to the working fluid.

ii. Adiabatic section (optional): space between evaporator and condenser, where no heat or cooling is applied.

iii. Condenser, where the vapour from the evaporator section of thermosyphon heat pipe is condensed usually by cooling water flowing through a water jacket.

iv. Insulation: the evaporator section is insulated to minimize heat losses.

The materials for the manufacturing of thermosyphon are carefully selected to ensure its effective performance. Other considerations are the type and the quantity of working fluid to be charged into the pipe.

2.2 Operation of thermosyphon heat pipe

The working principles of thermosyphon are similar to that of the wick heat pipe, but differ in the process of the return of the condensed liquid in the condenser due to the absence of wick structure. For proper operation of thermosyphon, the condenser is placed at the top of the evaporator so that the condensed liquid will return to the evaporator by gravity. **Figures 4** and **5** show a schematic diagram and a model of a typical thermosyphon (constructed in the University of Birmingham, UK) with heat supplied by coil of wire and heat rejected to the flowing water in the water jacket provided on the condenser section [17]. However, in some operation set ups, the heat can be supplied by hot water surrounding the evaporator of the pipe.

2.2.1 Operation limits of heat pipe

Heat pipe (with or without wick materials) operates within certain limits which are shown in **Figure 6**. For the heat pipe to operate, the maximum capillary pumping pressure must be greater than the total pressure drop; thus:

$$\Delta P_{c,\max} \geq \Delta P_l + \Delta P_v + \Delta P_g \tag{1}$$

The pressure drop is the sum of the following:

ΔP_l = Pressure drop necessary for the liquid to return from the condenser to the evaporator.

ΔP_v = Pressure drop necessary for the vapour to rise from the evaporator to the condenser.

ΔP_g = Pressure due to gravity whose value depends on the angle of inclination of the pipe.

If condition in Eq. (1) is not met (capillary limit), then the wick materials will dry out and the pipe will not operate. Detailed discussions on the heat pipe limits (shown in **Figure 6**) are available in heat pipe books, which can be referred.

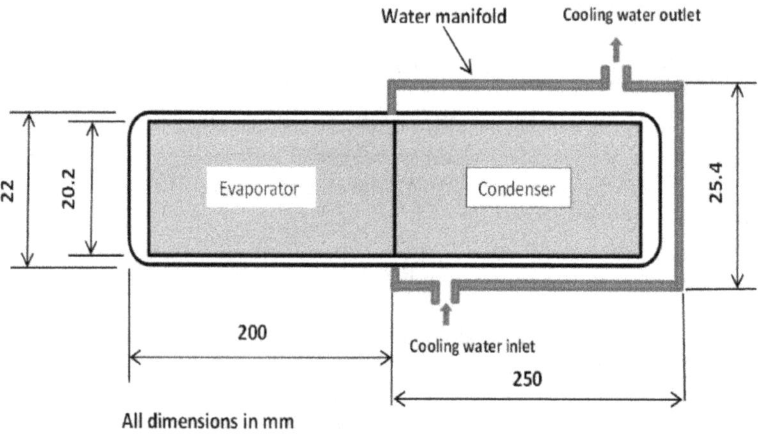

Figure 4.
Dimensions of a typical thermosyphon with water manifold [17].

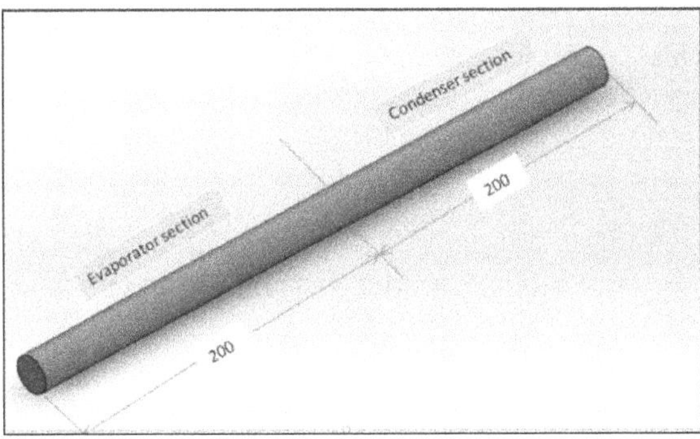

Figure 5.
3D view of a typical thermosyphon pipe.

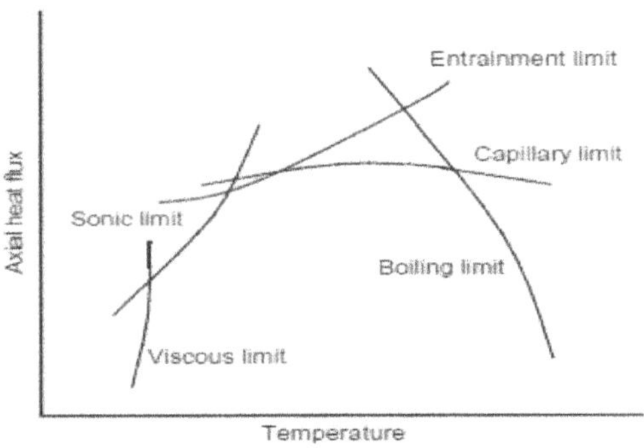

Figure 6.
Limitation of heat pipe for heat transport.

2.3 Advantages of thermosyphon over wick heat pipe

Apart from the general advantages of heat pipe, thermosyphon has other advantages over wick heat pipe, some of which are listed below:

i. Relative low-temperature difference between the heat source and heat sink

ii. More compactness

iii. High durability and reliability

iv. Cost-effectiveness

v. Less weight due to the absence of wick materials

vi. Simplicity in construction

2.4 Measurement of the performance of thermosyphon

The performance of thermosyphon under different conditions is evaluated based on the overall thermal resistance R_{th}, given by:

$$R_{th} = \frac{T_{ae} - T_{ac}}{Q_{in}} \tag{2}$$

where T_{ae} and T_{ac} are respectively the average temperatures on the evaporator and condenser while Q_{in} is the heat supplied to the evaporator.

However, the performance of the thermosyphon can also be calculated as the ratio of the heat transfer to the cooling water to the heat input as [18]:

$$\eta = Q_{out} / Q_{in} \tag{3}$$

The rate of heat transfer to the cooling water, Q_{out}, can be evaluated by:

$$Q_{out} = \dot{m} C_p (T_{out} - T_{in}) \tag{4}$$

where T_{in} and T_{out} are respectively the inlet and outlet temperatures of the cooling water, while $m\cdot$ and C_p are the mass flow rate, kg/s and the specific heat capacity of water, kJ/kg-K respectively.

Two approaches are usually employed in the performance characterization of thermosyphon, namely:

- Experimental
- Numerical

2.4.1 Experimental study on the performance of thermosyphon

The thermosyphon heat pipe can be experimentally characterized and the effects of some parameters on its performance evaluated. **Figures** 7 and 8 show a schematic diagram and picture of a typical test rig for the performance characterization of thermosyphon constructed at the University of Birmingham, UK, for analyzing the performance of a two-phase closed thermosyphon. It consists of a 0.4-m-long two-phase closed thermosyphon heat pipe, heating coil, water jacket and other instrumentations.

The heat can be supplied by hot water circulating around the evaporator or by electric power supply. In **Figures 6** and 7, the evaporator section is wrapped evenly with electric wire with electric energy supplied and controlled by TSx1820P Programmable DC PSU 18 V/20A power regulator to provide the heat required for boiling the working fluid inside the pipe. A multimeter is used for measuring the voltage input which is connected close to the pipe to account for the voltage drop

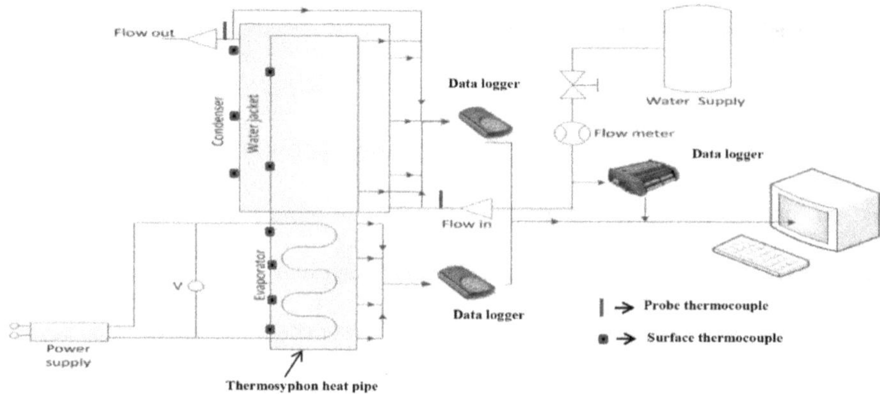

Figure 7.
Schematic diagram of the experimental test rig for thermosyphon characterization [17, 19].

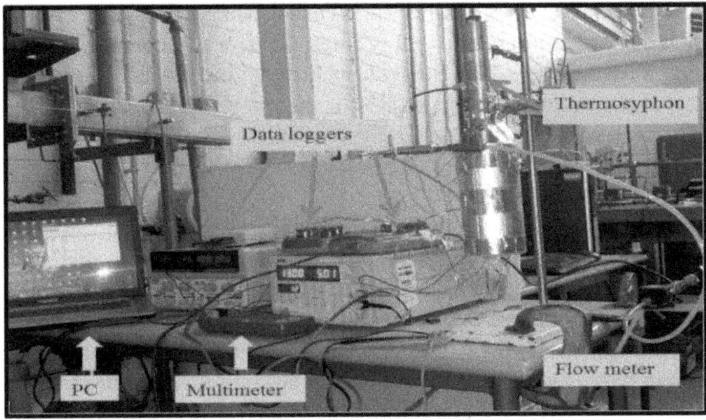

Figure 8.
Picture of the heat transfer characterization of thermosyphon test rig [17, 19].

while the current was read from the power regulator. The evaporator section is also insulated with 25-mm-thick pipe insulator to reduce the heat loss to the ambient environment (**Figure 8**). For measuring the temperature distribution along the pipe, 12 surface thermocouples were placed at different locations on the test pipe; 4 on the evaporator wall (at 0.02, 0.07, 0.12 and 0.17 m from the tip of the evaporator) and 2 on the condenser wall at 0.25 and 0.35 m as shown in the figures. The electric wires were wrapped in such way that they are not directly on the thermocouples so as to not affect their readings. Two probe thermocouples were installed at the inlet and outlet of the manifold to measure the temperatures of the cooling water. Three other thermocouples were used on the water jacket and one on the insulator to measure the effectiveness of the insulation and the jacket. All the readings were sent to Pico TC-08 data loggers connected to a PC.

2.4.1.1 Instrumentation and calibration

The test rig has to be provided with different measuring devices of temperature, water flow rate, heat (power) input and angular orientation to enable investigating the flow and heat transfer characteristics of the selected thermosyphon. The instruments include:

- ☐ Thermocouples, both surface and probe types.
- ☐ Flow meter.
- ☐ Electric power regulator (or hot water supply in some cases).
- ☐ Data logger.
- ☐ Angular measurement instrument such as protractor

The instruments are calibrated against standard devices and error analysis and uncertainties of their measurements are evaluated.

2.4.1.2 Experimental procedure

The test facility was completed and ready for investigations when all the parts were connected and water circulation system was checked for possible leakages. The operating conditions are set based on the type of the investigation to be carried out. However, in all the cases, the system is allowed to run and stabilize before readings are taken. Preliminary tests are required to determine the time when the system reaches steady state. Certain number of readings are set to be taken for each boundary condition at a set interval of time (usually in seconds). The reading recorded includes the temperatures, flow rates, voltage and current. Various investigations can be carried out using the test rig such as the effects of heat inputs, cooling water flow rate, inclination effects of the pipe, fill ratio, etc. Detailed procedure for each case depends on the type of the investigation to be carried out.

2.4.2 Numerical approach

To enable several investigations on many parameters affecting the performance of thermosyphon with different boundary conditions, numerical approach is usually employed. This is because experimental approach requires more time, energy and huge investment, to investigate many cases under different boundary conditions. There are two numerical approaches that are employed in modelling multiphase flows, namely the Euler-Euler and Euler-Lagrange approaches. In the Euler-Euler approach, the several phases are considered as interpenetrating continua mathematically in which each phase a volume is occupied only without sharing with other phases, while Euler-Lagrange approach utilizes Navier-Stokes equations that are solved for the fluid phase with several numbers of particles tracked in order to solve the dispersed phase. It should be noted that this approach cannot be adopted for applications in which volume fraction is important, especially for the secondary phase. Hence, the Euler-Euler approach is usually used in modelling two-phase closed thermosyphon operations.

Using Euler-Euler approach, three multiphase models are available in ANSYS Fluent:

a. The Eulerian model

b. The Mixture model

c. The Volume of Fluid (VOF) model

The mixture model deals with modelling of sedimentation, bubbly flows, particle-laden flows, etc. While applications such as fluidized beds, particle suspension, risers are modelled using Eulerian approach, on the other hand, liquid-gas tracking under steady or transient, free-surface flows, large bubble in liquid are modelled using the VOF approach.

Numerical modelling like computational fluid dynamic analysis (CFD) is an alternative to experimental approach, whereby several studies can be carried out with small investment. In CFD, a set of discretized equations are solved with the help of computer to get an approximate solution [20]. CFD analysis can be carried out on the flow and heat transfer characteristics of a thermosyphon heat pipe in both vertical and inclined orientations using a commercial ANSYS Fluent or any software that can model the simultaneous evaporation and condensation processes taking place in a thermosyphon heat pipe. However, some approaches like volume of fluid (VOF) in ANSYS Fluent require the user to add a user-defined function (UDF) to the modelling process.

The first step in solving any multiphase problem is identifying the suitable multiphase regime which represents the flow needed to be modelled. In this chapter, emphases is put more on the VOF model.

2.4.2.1 Model building

For building a model for simulating the flow and heat transfer characteristics of thermosyphon, a researcher is required to have a good knowledge of the theory (physics) behind the processes. The processes involved in the CFD modelling of the performance of thermosyphon using volume of fluid (VOF) approach in ANSYS Fluent can be summarized as follows:

i. Generation of the pipe geometry (model).

ii. Meshing of the model: different meshes of different properties (number of cells, faces, quality, etc.) are required.

iii. Carrying out a grid independence test: this is done to find out the situation whereby the result is independent of the mesh configuration and to select the configuration which will give less computational time.

iv. Importing the selected meshed file for the investigations into the ANSYS Fluent.

v. Attaching the user-defined function (UDF); this depends on the modelling approach selected.

vi. Modelling and simulation set up, which includes.

- Defining the boundary conditions.
- Setting the thermophysical properties of the materials involved such as thermal conductivity, material properties, density, specific heat capacity, viscosity, etc.
- Defining of the solution method and convergence.
- Running the simulation and processing of the results.
- Validation of the model: to enable validation of the developed model, the boundary conditions and other definitions are made exactly as those set in the experiment.
- Once the model is validated with the experimental results, then it can be used for further investigations.

2.5 Factors affecting the operations of thermosyphon

Considerable experimental research works were published on the investigation of the effects of parameters like the geometry, working fluid, fill factor and inclination on the thermosyphon heat pipe performance [21–25]. Hence, apart from the material of the thermosyphon, other important parameters affect its performance, such as:

I. Type of working fluid charged: The common liquid used in thermosyphon is water due to its availability, low cost, safety, etc. Below are some of the prime requirements for a liquid to be used in heat pipe:

i. Compatibility with wick and wall materials

ii. Good thermal stability

iii. Wettability of wick and wall materials: it is necessary for the working fluid to wet the wick and the container material, that is contact angle should be zero or very small

iv. High latent heat: a high latent heat of vaporisation is desirable in order to transfer large amounts of heat with minimum fluid flow, and hence to maintain low pressure drops within the heat pipe

v. High thermal conductivity: the thermal conductivity of the working fluid should preferably be high in order to minimize the radial temperature gradient and to reduce the possibility of nucleate boiling at the wick or wall surface

vi. Low liquid and vapour viscosities: the resistance to fluid flow will be minimized by choosing fluids with low values of vapor and liquid viscosities

vii. High surface tension: in heat pipe design, a high value of surface tension is desirable in order to enable the heat pipe to operate against gravity and to generate a high capillary driving force

viii. Acceptable freezing or pour point

The selection of the working fluid must be based on thermodynamic considerations which are concerned with the various limitations to heat flow occurring within the heat pipe, like viscous, sonic, capillary, entrainment and nucleate boiling levels.

Some common liquids used in heat pipe include water, acetone, ethanol, ammonia, nitrogen and methanol. However, recent researches have shown potentials of using other liquids alone or mixed with water like nanofluids [26–28].

II. Quantity of the working fluid charged: the quantity of the liquid charged in relation to the volume of the evaporator, called fill ratio, FR or liquid ratio, plays a vital role in the performance of thermosyphon. Fill ratio is defined as the ratio of volume of the working fluid in an unheated pipe, V_{liq}, to the volume of the evaporator, V_e:

$$FR = {}^{V_{liq}}\!/_{V_e} = \frac{4V_{liq}}{\pi D^2 l_e} \tag{5}$$

The quantity of the fluid to be charged has to be properly selected, which depends on the intended applications, as insufficient amount of fluid causes dry out while excessive amount reduces performance and increases the cost of the pipe. FR of a thermosyphon should be between 40 and 60% for vertical pipes and between 60 and 80% for inclined pipes [4, 29] . For example, Emami et al. [30] and Asgar [18] obtained 45 and 50% as best FR respectively.

III. Heat input: The amount of heat supplied in the evaporator affects the performance of the thermosyphon depending on other factors such as size, fill ratio, its geometry and operating limits. Experimental results have shown that the performance of the thermosyphon increases with the increase in heat input up to their operating limits. It increases with increase between 350 and

500 W, but it decreases when the heat input is above 500 W [18] . But for Abdullahi et al. [19], the performance of the pipe increases as the heat input increases from 20 to 81.69 W, but it tends to decrease as more heat is supplied, showing the limit of this pipe has been reached under these operating conditions (**Figure 9**). Hence, the trend of the performance of the thermosyphon (based on the amount of the heat input in the evaporator section) depends on its operating limits. At low heat input, the vapour generated from the evaporator section is small, so there will be significant dry areas in the condenser section; hence, heat transfer is largely by free convection. As the heat is gradually increased, more vapour will rise to the condenser section, there will be high condensation rate on the condenser wall and the dominant heat transfer mechanism will be condensation. But at certain high heat input, thick layer of liquid can be formed on the wall of the pipe causing high thermal resistance and hence lower the heat transfer to the cooling water, hence reduction of performance.

IV. Inclination angle: since the condenser of thermosyphon must be at the top with the evaporator at the bottom for the condensate to return, this shows that the pipe can be inclined at any angle other than 90°. Regarding the effect of inclination angle on heat pipe performance, conflicting experimental results were reported like angles between 15 and 60° [24], between 40 and 45° [25] and 60° [30] gave the best performance. Others reported higher angles like 90° [31] and 83° [32] as the best performing angles while few reported that inclination angle has no effect [33]. The possible reasons for the contradicting results are the complex nature of the processes taking place in thermosyphon operations and various parameters affecting its performance. Furthermore, those researches are only experimental and considered a small range of inclination angles. With the contradictory experimental results in the literature and lack of, or limited, numerical studies on the effect of inclination, Abdullahi et al. [19] addressed these issues through the development of a CFD model that studied the effects of inclination angles (10–90°) and experimentally validated the model. Experimental and numerical results showed that increasing the inclination angle will improve the thermosyphon heat pipe performance to reach its maximum value at 90°, but this effect decreases as the heat input increases [19] (**Figure 10**).

V. Flow rate of cooling water: the rate at which cooling water is passing in the water jacket around the condenser of a thermosyphon affects its performance.

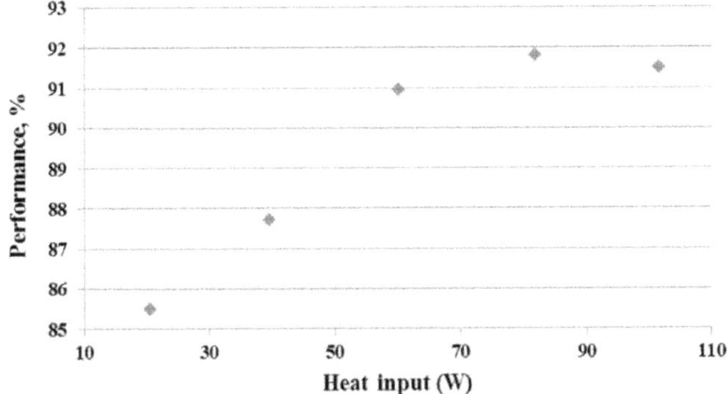

Figure 9.
Performance of thermosyphon aligned vertically at different heat inputs [17, 19].

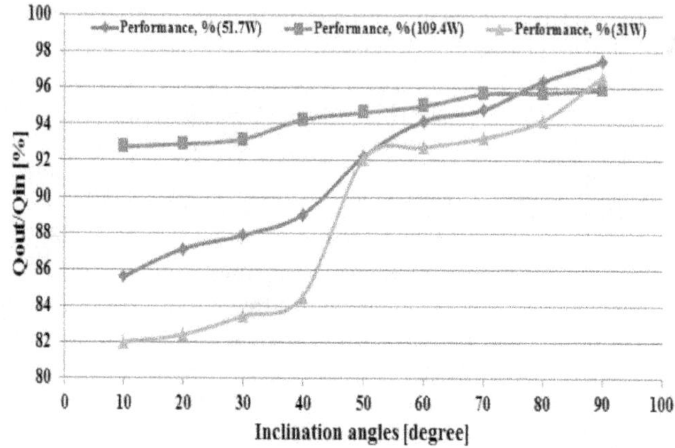

Figure 10.
Variation of the thermosyphon performance with inclination angle at different heat inputs [17, 19].

As the rate of the heat removal from the vapour increases, more condensate returns to the evaporator for another cycle. The effect of cooling water flow rate at constant heat input was investigated on the performance of thermosyphon heat pipe [19]. The heat input was fixed at 101 W while five different flow rates ranging from 0.00156 to 0.00611 kg/s were investigated. Temperature and the flow rate readings were recorded for each run and the effects of the cooling water flow rate were evaluated based on the overall thermal resistance, rate of heat transfer to the cooling water, outlet temperature of cooling water, performance of the thermosyphon, etc. The results from such work have shown that the performance of the pipe in terms of heat transfer to the cooling water increases with the increase in the cooling water flow rate. This is due to the mass flow of the cooling water which results in the enhancement of the rate of heat transfer from the pipe wall to the cooling water and subsequent increase in the efficiency.

2.6 Applications of thermosyphon

In addition to the general advantages of heat pipes, thermosyphon type is found to be highly durable, reliable and cost-effective, which make them useful for various applications, such as:

I. Solar heating of building [16].

II. Liquid circulation: thermosyphon system is used for circulating liquids and volatile gases in heating and cooling systems such as water heaters, furnaces and boilers. It simplifies transfer of liquid or gas without using conventional pump which adds cost and complexity to the system.

III. Cooling applications: thermosyphon is used in cooling of turbine blades, transformers, electronics, internal combustion engines and nuclear reactors [34, 35]. This is due to their ability to dissipate and transfer large amount of energy from small area without any significant loss.

IV. Aircraft cooling: due to their light weight, thermosyphon pipes are used in cooling of aircraft and spacecraft.

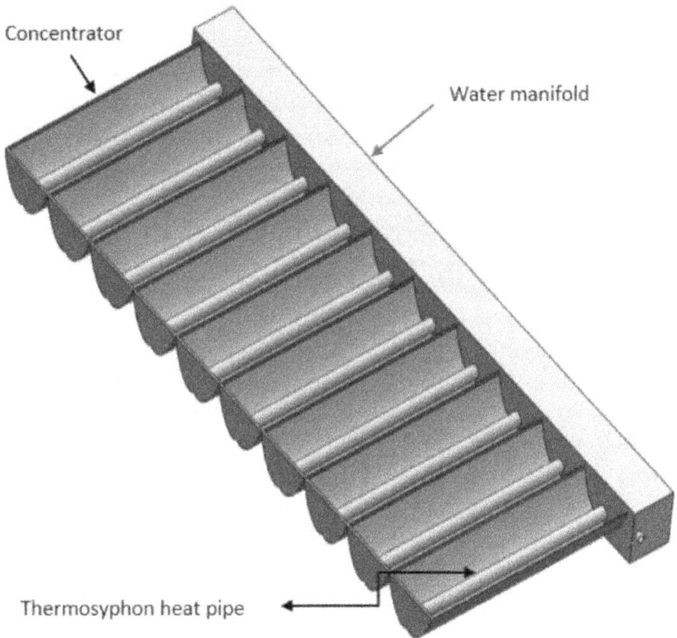

Figure 11.
Developed compound parabolic collector with thermosyphon as receiver [17].

Receiver in solar collector (solar systems): thermosyphon is proved to be a good choice as a receiver for solar concentration systems due to its advantages stated [36, 37] as shown in **Figures 3** and **11**.

3. Conclusions

Several parameters affect the operation of thermosyphon such as fill ratio, working fluid, inclination, geometry, heat input, cooling water flow rate, etc. Experimental and numerical (CFD) studies are usually carried out to enable the investigation of the effects of some of these parameters on the performance of thermosyphon heat pipe for use in various engineering applications. Investigations on the effects of heat input, fill ratio, flow rate of cooling water on the temperature distributions on the wall of the pipe, overall thermal resistance and overall performance of the pipe at vertical orientation were shown to be possible both experimentally and using CFD. Also, the effect of inclination angle of thermosyphon on those parameters was successfully added in the Fluent. Hence, the chapter has shown that volume of fluid (VOF) model's approach in ANSYS together with UDF and other software can fully simulate the complex evaporation and condensation processes taking place in thermosyphon for both vertical and inclined orientations.

Author details

Bala Abdullahi[1]*, Raya K. Al-dadah[2] and Sa'ad Mahmoud[2]

1 Department of Mechanical Engineering, Kano University of Science and Technology (KUST), Wudil, Kano, Nigeria

2 School of Mechanical and Manufacturing Engineering, University of Birmingham, United Kingdom

*Address all correspondence to: balabdullahi@yahoo.com

References

[1] Reay DA, Kew PA. 7—Applications of the Heat Pipe. In: Reay DA, Kew PA, editors. Heat Pipes: Theory, Design and Applications. 6th ed. Oxford: Butterworth-Heinemann; 2007. pp. 275-317

[2] Joudi KA, Witwit AM. Improvements of gravity assisted wickless heat pipes. Energy Conversion and Management. 2000;**41**(18): 2041-2061

[3] Peterson GP. An Introduction to Heat Pipe, Modelling, Testing and Applications. New York, USA: John Wiley and Sons Inc; 1994

[4] ESDU 80013. Heat Pipes—General Information on Their Use, Operation and Design. ESDU: ESDU International PLC; 1980

[5] Reay D, Ryan M, Peter K. Heat Pipes: Theory, Design and Applications. Oxford, USA: Butterworth-Heneman; 2014

[6] Deverall JE, Kemme JE. Satellite Heat Pipe. University of Califonia: Los Abamos Scientific Laboratory; 1970

[7] Annad DK. Heat Pipe Application to a Gravity Gradient Satellite. In: ASME Annual Aviation and Space, Baverley Hills, CAL, 1968

[8] Vasiliev LL. Heat pipes in modern heat exchangers. Applied Thermal Engineering. 2005;**25**(1):1-19

[9] Chow TT, He W, Ji J. Hybrid photovoltaic-thermosyphon water heating system for residential application. Solar Energy. 2006;**80**(3): 298-306

[10] Abreu SL, Colle S. An experimental study of two-phase closed thermosyphons for compact solar domestic hot-water systems. Solar Energy. 2004;**76**(1–3):141-145

[11] Wei L, Yuan D, Tang D, Wu B. A study on a flat-plate type of solar heat collector with an integrated heat pipe. Solar Energy. 2013;**97**:19-25

[12] Pastukhov VG et al. Miniature loop heat pipes for electronics cooling. Applied Thermal Engineering. 2003; **23**(9):1125-1135

[13] Vasiliev LL. Micro and miniature heat pipes—Electronic component coolers. Applied Thermal Engineering. 2008;**28**(4):266-273

[14] Chami N, Zoughaib A. Modeling natural convection in a pitched thermosyphon system in building roofs and experimental validation using particle image velocimetry. Energy and Buildings. 2010;**42**(8):1267-1274

[15] Wonston S, Kiatsiriroat T. Effects of inclined heat transfer rate on thermosyphon heat pipe under sound wave. Asian Journal on Energy and Environment. 2009;**10**:214-220

[16] Pouland ME, Fung A. Potential benefits from thermosyphon—PCM (TP) integrated design for buildings applications In Toronto. in Canadian conference on building simulation (eSim 2012); Toronto: 2012

[17] Abdullahi B. Development and Optimization of Heat Pipe based of Compound Parabolic Collector. UK: School of Mechanical Engineering, University of Birmingham; 2015. p. 287

[18] Asghar A, Masoud R, Ammar AA. CFD modeling of flow and heat transfer in a thermosyphon. International Communications in Heat and Mass Transfer. 2010;**37**:312-318

[19] Abdullahi B, El-sayed A, Al-Dadah RK, Mahmoud S, Abdel Fateh M, Muhammad NM, et al. Experimental

and numerical investigation of thermosyphon heat pipe performance at various inclination angles. Journal of Advanced Research in Fluid Mechanics and Thermal Sciences. 2018;**44**(1):85-98

[20] Tannehill J, Anderson DA, Pletcher RH. Computational Fluid Mechanics and Heat Transfer. 2nd ed. US: Taylor and Francis; 1997

[21] Gabriela H, Angel H. Heat transfer characteristics of a two—phase closed thermosyphons using nanofluids. Experimental Thermal and Fluid Science. 2011;**35**:550-557

[22] Jouhara H, Ajji Z, Koudsi Y, Ezzuddin H, Mousa N. Experimental investigation of an inclined—condenser wickless heat pipe charged with water and ethanol—water azeotropic mixture. Energy. 2013;**61**:139-147

[23] Chehade AA, Louahlia-Gualos H, Masson SL, Voicu I, Abouzahab-Damaj N. Experimental investigation of thermosyphon loop thermal performance. Energy Conversion and Management. 2014;**84**:671-680

[24] Noie SH, Emami MRS, Khoshnoodi M. Effect of inclination angle and filling ratio on thermal performance of a two-phase closed thermosyphon under normal operating conditions. Heat Transfer Engineering. 2007;**28**(4): 365-371

[25] Hahne E, Gross U. The influence of the inclination angle on the performance of a closed two-phase thermosyphon. Journal of Heat Recovery Systems. 1981;**1**(4):267-274

[26] Mathioulakis E, Belessiotis V. A new heat-pipe type solar domestic hot water system. Solar Energy. 2002;**72**(1):13-20

[27] Arab M, Soltanieh M, Shafii MB. Experimental investigation of extra-long pulsating heat pipe application in solar water heaters. Experimental Thermal and Fluid Science. 2012;**42**:6-15

[28] Aung NZ, Li S. Numerical investigation on effect of riser diameter and inclination on system parameters in a two-phase closed loop thermosyphon solar water heater. Energy Conversion and Management. 2013;**75**:25-35

[29] Nguyen-Chi H, Groll M. Entrainment or Flooding limit in a closed two—phase thermosyphon. In Advances in Heat Pipe Technology: IV International Heat Pipe Conference; Oxford, London: Pergamon Press; 1982

[30] Emami MRS, Noie SH, khoshnoodi M. Effect of aspect ratio and filling ratio on the thermal performance of an inclined two—phase closed thermosyphon. Iranian Journal of Science and Technology. 2008;**32**:39-51

[31] Karthikeyan H, Vaidyanthan S, Sivaraman B. Thermal performance of a two-phase closed thermosyphon using aqueous solution. International Journal of Engineering Science and Techonolgy. 2010;**2**(5):913-918

[32] Grooten MHM, Vander Geld OWM. Effects of angle of inclination on the operation limiting heat flux of long R-134a filled thermosyphons. Journal of Heat Transfer. 2010;(5):**132**

[33] Ong KS, Tong WL. Inclination and fill ratio effects on water filled two phase closed thermosyphon using a straight grooved and helical grooved. In: International Heat Pipe Symposium (IHP); Taipei, Taiwania: 2011

[34] Japikse D. Advances in thermosyphon technology. Advances in Heat Transfer. 1973;**9**

[35] Mochizuki M et al. Nuclear reactor must need heat pipe for cooling. In: International Heat Pipe Symposium (IHPS). Taipei, Taiwania; 2011

[36] Jouhara H, Chauhan A, Nannou T, Almahmoud S, Delpech B, Wrobel LC. Heat pipe based systems—Advances and applications. Energy. 2017;**128**: 729-754

[37] Grissa K, Benselama AM, Romestant C, Bertin Y, Grissa K, Lataoui Z, et al. Performance of a cylindrical wicked heat pipe used in solar collectors: Numerical approach with Lattice Boltzmann method. Energy Conversion and Management. 2017;**150**:623-636

Chapter 3

The Recent Research of Loop Heat Pipe

Chunsheng Guo

Abstract

The loop heat pipe was first studied for the difficult temperature control conditions under aerospace conditions. The loop heat pipe is composed of evaporator, reservoir, capillary wick, vapor/liquid line, and condenser. Different working fluids, different liquid filling amounts, different capillary wicks, different sizes, and different cooling methods will have an important impact on the performance of the loop heat pipe. Therefore, if the loop heat pipe wants to have good heat transfer efficiency, it is imperative to discuss good processing steps and processing techniques. When the loop heat pipe is running, the capillary wick is heated, the liquid in the capillary wick is heated and vaporized, and the gas passes through the vapor line to enter the condenser for condensation. After the condensation, the liquid flows back into the reservoir and the inside of the capillary wick through the liquid line. How to ensure the forward operation of the gas at the evaporator to reduce the reverse leakage heat during this cycle, how to ensure the condenser condensation efficiency is sufficient, etc. are all issues to be considered. This chapter describes in detail the processing and optimization methods for each part, and prepares a loop heat pipe that can work normally.

Keywords: nickel-ammonia loop heat pipe, capillary wick, heating power, temperature fluctuation

1. Introduction

The loop heat pipe is a variant of the ordinary heat pipe, and is a two-phase flow loop type heat pipe, which is a unidirectional heat conduction element similar to a diode. The working medium is mainly composed of a gas phase and a liquid phase in the steam pipe and the liquid pipe respectively. The form of the cycle, complete the transfer of heat. The design idea of the world's first LHP system was proposed by scientists in 1971. Then Soviet scientists Maydanik and Gerasimov successfully developed the world's first LHP system in 1972, after the Soviet Union and the United States. Research institutions are beginning to study loop heat pipes. The loop heat pipe is a high-efficiency loop heat transfer device with two-phase separation heat transfer. Since the structure of the loop heat pipe is more complicated than the ordinary heat pipe, it can adapt to more different working environments. The loop heat pipe has many advantages such as high heat transfer performance, long-distance heat transfer, small heat transfer temperature difference, and flexible installation. Therefore, in recent years, it has been widely used in aerospace heat dissipation, electronic products heat dissipation, anti-gravity ground heat transfer working environment and other fields.

Loop heat pipe was a new kind of two phase flow heat equipment, it used capillary suction force drive the working medium to complete the cycle of working medium inside the heat pipe flow and by using phase change of working medium two-phase flow to transfer heat [1]. Loop heat pipe had many advantages, such as large heat transmission, transmission distance, high heat transfer efficiency, antigravity features was strong, etc. [2]. Loop heat pipe was first applied in aerospace field, with the constant improvement of the loop heat pipe technology, it gradually be applied to every other civilian areas, especially in electronic cooling field [3].

The most important characteristic of loop heat pipe was its resistance to gravity characteristic. Baumann [4] made a theoretical analysis first for the influence of loop heat pipe resistance to gravity, but fails to provide the experimental data to support. Zhang [5] studied the loop heat pipe in start-up and heat transfer performance under antigravity work condition, found the loop heat pipe started up under the condition of antigravity liquid reflux need to overcome additional pressure drop loss gravity bringing, and in outside loop resistance increased obviously. When vapor trough exist vapor, start-up time and temperature increase, loop heat pipe appeared complex compound startup phenomenon. Anti-gravity work increased the working temperature of the loop heat pipe system, reduced the automatic temperature control range, and leaded to vapor produced in the evaporator was more likely to overheat and overall system thermal resistance increased at the same time. In order to match the cooling parts, the evaporators were often made into plates [1]. Research team of Chinese Academy of Sciences institute of physics [6] carried out the establishment of the numerical model of the plate LHP evaporator and the heat transfer model, then they analyzed the heat transfer mechanism inside the evaporator. Mitomi [7] designed three kinds of length loop heat pipe with the length were 2, 5, 10 m. Loop heat pipe working medium was ethanol and capillary core was made by polytetrafluoroethylene porous material. They analyzed the normal working performance and carried out numerical simulation, and found that three pipes could work under the same starting power. South China University of Technology Developers [8] had studied the influence of heat leakage, the initiation characteristics and optimized the evaporator, reservoir and condenser. Beijing Aerospace University and Institute of Space Studied [9, 10] studied the whole loop heat pipe system, including the evaporator and capillary core structure optimization effect on the performance of the loop heat pipe, the influence of reservoir auxiliary cooling and the evaporator auxiliary heat on loop heat pipe performance, performance test of loop heat pipe in microgravity environment, the influence of quantity of quality filling on the performance of the loop heat pipe. In practical application, there could be multiple sources of heat and multiple sources. The number of evaporators, condensers and accumulators is not fixed. Jentung Ku [11] tests the multi-evaporator loop heat pipe with 50 W power in a vacuum environment, and the performance is stable.

2. Capillary core preparation

Capillary core is an important component in loop heat pipe. Porosity, permeability, pore size distribution and capillary suction ability are the key parameters to reflect the performance of capillary core. In order to improve the ability of anti-gravity operation and long-distance operation of the loop heat pipe, it is necessary for the capillary core to have the characteristics of high permeability and high capillary suction ability. As the core device in the loop heat pipe, the capillary core transmits heat in the evaporator and provides enough capillary force to drive the working fluid cycle. At the same time, the vapor should be transferred to the vapor pipe in time to prevent the phenomenon of suction. Wolf indicates that LHP has gravity heat pipe and capillary pump heat pipe. It has the advantages of anti-gravity, long distance operation, no external power source, high stability, passive energy transportation and so on. Capillary core, as the core component of loop heat pipe, provides the necessary power for the forward operation of heat pipe. At present, the main types of capillary core are: grooved capillary core, metal mesh capillary core, ceramic capillary core and metal powder sintered capillary core. At present, most metal powder sintered capillary cores are made of copper and nickel. These capillary cores are widely used in loop heat pipes because of their good thermal properties and liquid compatibility.

On the basis of previous studies, double pore capillaries were prepared by molten salt pore making technique. Compared with the addition of volatile pore-making agent, NaCl, with 99.5% purity and carbonyl nickel powder as capillary core material has the advantages of easy removal and uniform void distribution. The main parameters of capillary core, such as porosity, permeability, thermal conductivity, pore distribution and capillary suction ability, were measured according to the ratio of pore-forming agent and the pressure of cold pressing molding. The effect of the ratio of pore-making agent and the pressure of cold pressing on the properties of capillary core was obtained.

2.1 Preparation process of capillary core

Nickel powder was selected as raw material and NaCl as pore-making agent to prepare double pore capillary core. The main steps were powder ratio, cold pressing, sintering and cleaning. The specific steps were as follows: the preparation process was shown in **Figure 1**.

1. Powder ratio: In this paper, 2 μm nickel powder with 99.5% purity of NaCl was selected as the material to prepare the double aperture capillary core. Firstly, the NaCl particles were ground by ball mill (the positive and negative rotation time was 45 min, the interval time was 5 min, the total milling time was 6 h), and the total milling time was 6 h, the positive and negative rotation time was 45 min, the interval time was 5 min, and the total milling time was 6 h. The particle size of NaCl was mainly distributed in 200–400 mesh after ball milling. There were very few NaCl particles below 400 mesh. The NaCl powder with diameter of 48 μm (300–400 mesh) was screened by vibrating screen, and then the nickel powder and NaCl powder were mixed evenly by ball mill, and then put into drying. The box was dried.

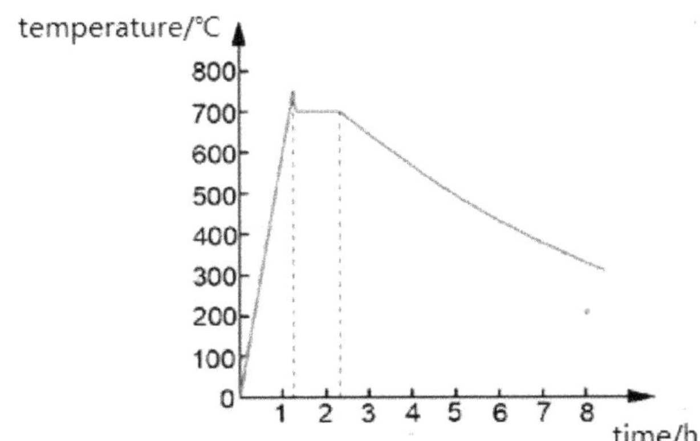

Figure 1.
Sintering temperature curve of nickel based capillary core.

2. Cold pressing molding. The powder was compacted by a press (the target pressure was 30, 40, 50, 60 KN, and the booster speed was 200 N/s).

3. The capillary core is sintered. The vacuum hot pressing sintering furnace selected in the experiment is ZT-40-20Y, combined with multiple sintering experiments and the temperature curve drawn in the previous literatures is shown in **Figure 2**.

4. Ultrasonic cleaning. After sintering, the NaCl particles in the capillary core need to be dissolved by ultrasonic cleaning to form a void and obtain a dual pore structure. The SEM of biporous wick is shown in **Figure 3**.

2.2 Capillary core parameter testing

2.2.1 Porosity and permeability

Porosity is the most direct index of porous structure of capillary core. The internal pores of capillary core are divided into connected pores, semi-connected pores, closed pores. The porosity of capillary core prepared by salt solution technique is mainly affected by the proportion and size of pore-forming agent.

Archimedes drainage method for porosity measurement, capillary core wet weight m_{wet}, drying thoroughly weighing capillary core dry weight m_{dry}, and measured the outer diameter of cylindrical capillary core r and length L, densities of deionized water. Porosity ε

$$\varepsilon = \frac{m_{wet} - m_{dry}}{V \cdot \rho} \times 100\% \tag{1}$$

According to the gas resistance test table shown in **Figure 4**, the experimental device uses compressed air as the air source, and the compressed air in the air compressor enters from the left side air compressor joint. In order to ensure the accuracy of the experiment and improve the service time of the platform, a filter and a steady pressure tank are connected after the air enters the pipeline. The gas flow through the experimental section is controlled by the mass flow-meter and the mass flow controller after the air passes through the unidirectional valve after the steady pressure. The capillary core is installed in the experiment section, and the pressure difference between the two sides of the capillary core is detected by pressure differential transmitter. At the end of the experiment, the gas was emptied into the air through the buffer tank.

The flow in this experiment is the flow inside the tube, the maximum flow rate is 30 L/min, the length of stainless steel tube is 300 mm, the inner diameter of stainless steel pipe is 20 mm, and the Reynolds number is

$$Re = \frac{\rho v d}{\mu} \tag{2}$$

The calculated Reynolds number is 176.9. The experimental fluid flow is laminar flow, which accords with the applicable condition of Darcy formula. The length of stainless steel tube is 15 times the length of capillary core. The working fluid in the pipe can be developed fully and the inlet effect can be reduced effectively. The thickness of stainless steel tube wall 1 mm is much smaller than the diameter of capillary core. This experimental device can accurately measure the permeability of porous media K. A number of empty tube experiments were carried out before the experiment to eliminate the pressure drop caused by friction on the pipe wall.

According to the results of many experiments by French scientist Darcy, permeability K can be measured experimentally by formula (3).

$$q_v = \frac{KA\Delta P}{\mu H} \tag{3}$$

q_v is the flow rate in the experiment (m^3/s), K is the permeability of the sample (m^2), A is the cross-sectional area of the sample (m^2), H is the length of the sample (m), ΔP is

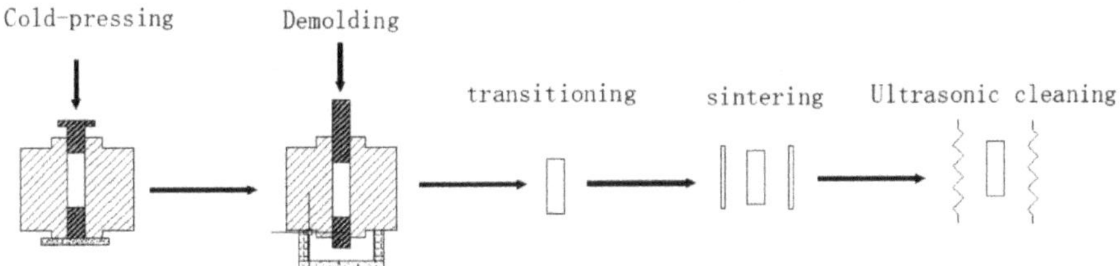

Figure 2.
Flow chart of capillary core preparation.

Figure 3.
50 kN cold pressure 20% NaCl mass fraction capillary core surface 1200 and 600 times scanning electron microscope.

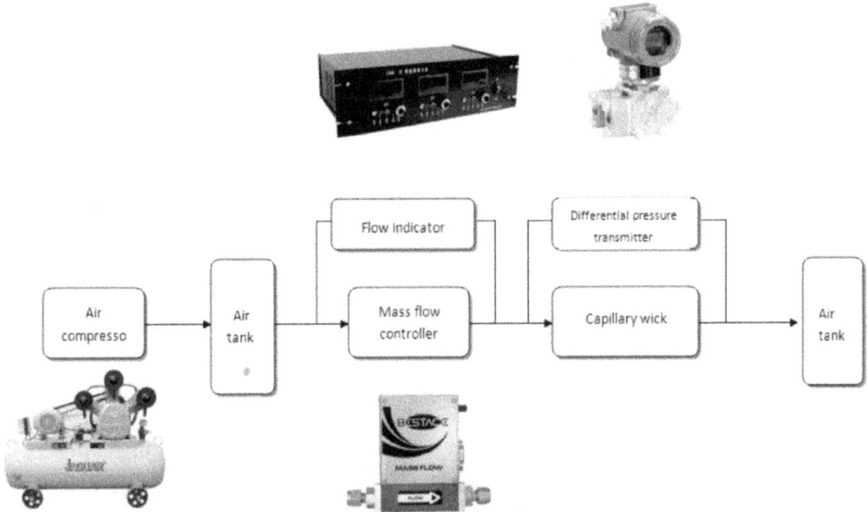

Figure 4.
Schematic diagram of gas resistance test table.

the pressure difference at the two ends of the sample (Pa), μ is experimental fluid viscosity (Pa s). The schematic diagram of the experimental device is shown in **Figure 4**. The principle is that the air compressor acts as a gas source to provide the experimental fluid for the whole experimental device, and the capillary core is put into the experimental section by controlling the flow rate of the experimental fluid by the mass flowmeter (q_v). According to formula (3), the permeability of capillary core is calculated by observing the pressure difference between the two ends of the experimental section by the pressure differential transmitter.

According to the experimental part, **Figure 5** shows the porous wicks porosity and permeability curves of different NaCl proportion.

It can be seen from the figure that as the proportion of NaCl increases, the porosity and permeability increase gradually. The reasons of this phenomenon are:

1. As the proportion of NaCl increases, the volume of NaCl particles increases, and the total powder mass is equal during cold press forming, so the porosity increases as the proportion of NaCl increases.

2. During the cold pressing and sintering process, as the temperature rises, the gap between the nickel powder particles gradually decreases, and the gap between the nickel powder and the NaCl particles remains unchanged, so the proportion of the pore former increases, and the porous wick shrinks. The smaller the degree, the larger the porosity.

3. The particle size of NaCl particles is significantly larger than the particle size of nickel powder. After cleaning and desalting, the original NaCl particles occupy the pores, the flow resistance of the working medium in the pores decreases, and the permeability increases.

Figure 6 is porous wick porosity and permeability curve for 20% NaCl wt of different cold forming pressures. It can be seen from the figure that as the cold forming pressure increases, the porosity remains basically unchanged and the permeability gradually decreases. The reasons are:

1. As the cold forming pressure increases, the pore size between the nickel powder particles decreases, and the working fluid flow resistance becomes larger, so the permeability decreases.

2. The proportion of total NaCl in the porous wick is the same. The pores occupied by NaCl particles account for the main part of the biporous structure. The proportion of pores formed between the nickel powder particles during sintering is small, so the porosity remains basically unchanged.

2.2.2 Capillary suction experiment

The relationship between capillary force and permeability is complex, the relationship between capillary force and permeability is negative, and the increase of permeability is bound to decrease. The most direct method for observing capillary force is to observe the liquid level rising velocity and suction mass velocity in porous media. In this paper, the suction ability of capillary core is determined by observing the suction quality of capillary core.

The principle of the experiment is that the bottom of the capillary core is in contact with the liquid surface by controlling the lifting platform to ensure that only

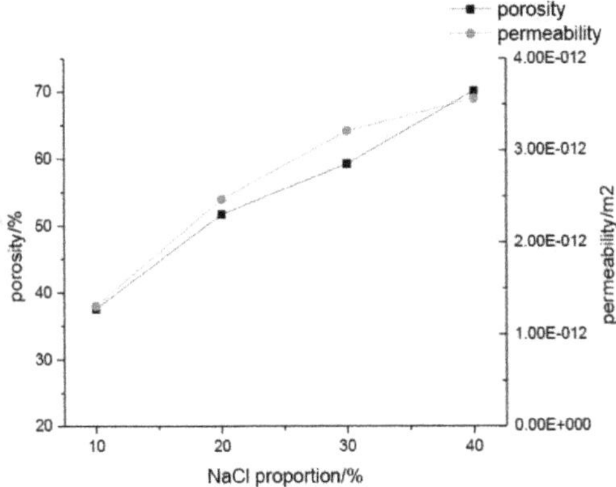

Figure 5.
Porosity and permeability curves of different NaCl proportion.

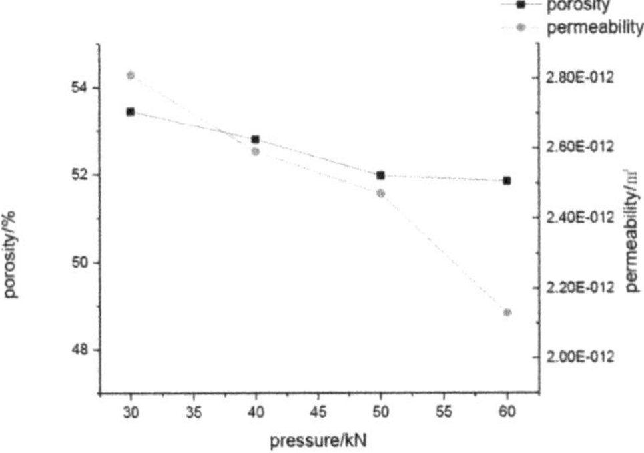

Figure 6.
Porosity and permeability curves of different cold pressing pressure.

the bottom of the capillary core exists the phenomenon of suction. The electronic analysis balance records the data and draws the suction curve.

As shown in **Figure 7**, the size of the capillary suction specimen is 100 mm in length and 20 mm in diameter. The suction fluid is deionized water, the bottom of the capillary core is in rigid contact with the deionized water surface through the motion control platform, and the quality of the working fluid inside the beaker is measured by the electronic analytical balance (accuracy is 0.0001 g). The reduced mass is the quality of capillary suction. In order to control the rising velocity of liquid level accurately, the minimum rising speed of motion control platform is 0.01 mm/s. In order to ensure the measuring accuracy and the total quality of suction fluid to reduce the influence of the inside wall of beaker on the suction process, the diameter of beaker is 40 mm. The height is 160 mm. **Figure 8** shows the physical model of capillary aspiration. The suction curve of different PFA agent ratio and different cold-forming pressure are shown in **Figure 9** and **Figure 10**.

Figure 11 (the number below the pore-forming agent ratio is the total number of pores in the range of 0–30 μm in the electron micrograph) is 80 times magnification of the surface of the porous wicks, and the surface pore size distribution is measured by image pro plus software, 10, 20, 30, 40%, respectively. It can be seen that most of the pore diameters are distributed at 2–4 μm (about 30%). It can be seen that the total pore size decreases in the range of 2–30 μm with the increase of the proportion of pore-forming agent, The reason is that the proportion of the

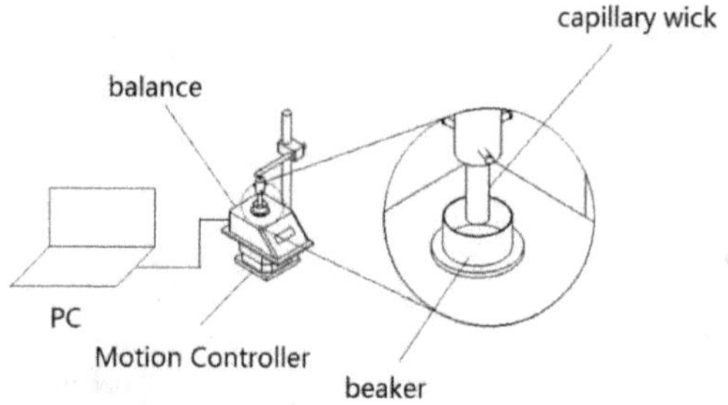

Figure 7.
Schematic diagram of capillary suction platform.

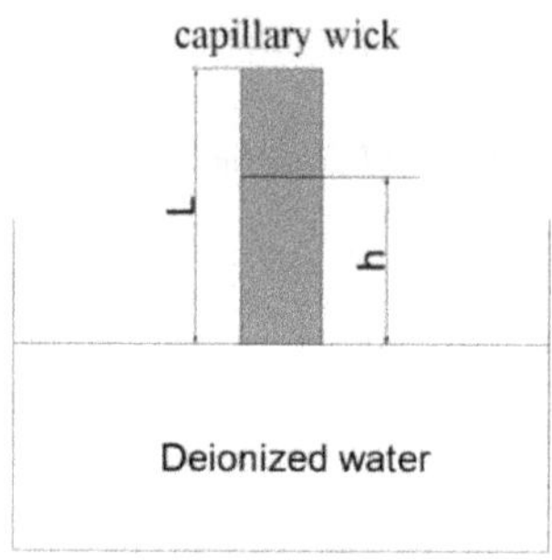

Figure 8.
Physical model of capillary aspiration.

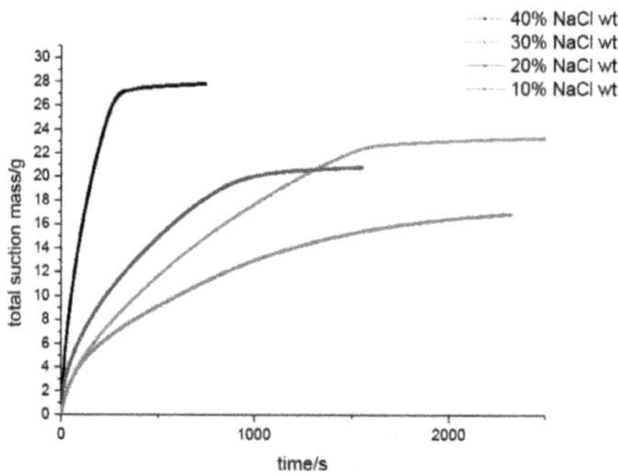

Figure 9.
The suction curve of different NaCl ratio.

pore-forming agent is increased, and the number of pore-forming agent particles in the same section is increased, and the pores of each size are adhered to each other, resulting in a decrease of the number of total pores.

As is shown in **Figure 10**, the capillary suction mass of the porous wick is proportional to the porosity, and the experimental results are in line with the derivation conclusion.

The capillary suction speed of the porous wick is proportional to the porosity and the average pore diameter. It can be seen from the figure that the porous wicks (40% NaCl wt) have the fastest capillary suction speed and the porous wicks (10% NaCl wt) have the slowest capillary suction speed. Porous wicks (20% NaCl wt) with a small diameter of pores more than porous wicks (30% NaCl wt), so the former has a higher suction speed than the latter.

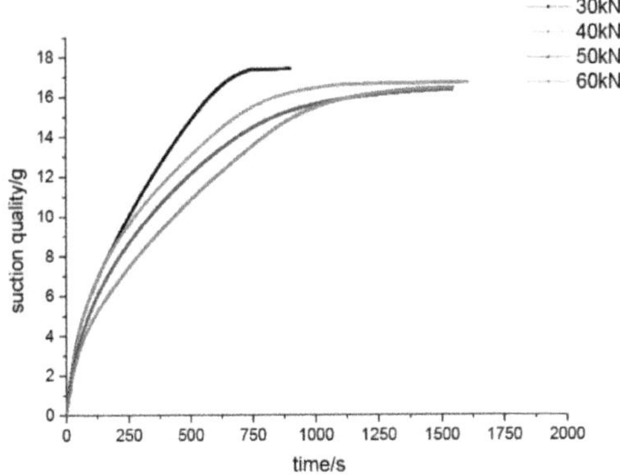

Figure 10.
The suction curve of different wicks with different cold pressure.

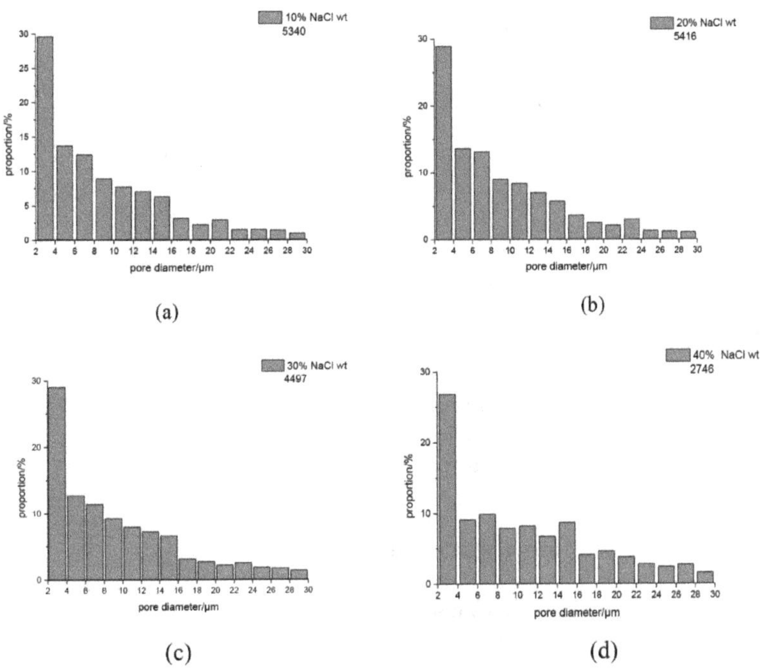

Figure 11.
Pore diameter distribution (a), (b), (c), and (d) are 10, 20, 30, and 40%, respectively.

As is shown in different cold forming pressures, suction speed porous wicks (30 kN) > porous wicks (40 kN) > porous wicks (50 kN) > porous wicks (60 kN), the suction quality is basically consistent but there are small differences, porous wick (30 kN) > porous wick (40 kN) > porous wick (50 kN) > porous wick (60 kN).

It can be seen that as the cold pressure increases, the porosity and permeability both decrease. When the length of the porous wick is controlled, the internal structure of the porous wick pressed at 30 kN pressure is loose, the permeability is large, the suction resistance is small, so the suction speed is faster, the permeability decreases gradually with the increase of pressure, the suction speed decreases. As is shown in **Figure 12** small pores which produce larger capillary force decrease

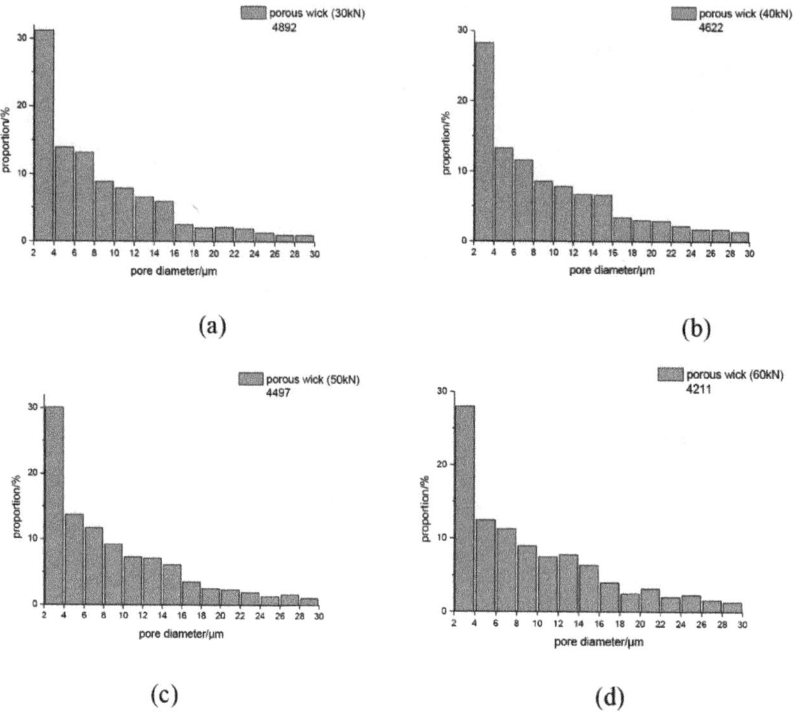

Figure 12.
Pore diameter distribution (a), (b), (c), and (d) are 30, 40, 50, and 60 kN, respectively.

with the increase of pressure, so the capillary suction speed is porous wicks (30 kN) > porous wicks (40 kN) > porous wicks (50 kN) > porous wicks (60 kN).

The total suction mass of the porous wicks with different cold forming pressures is porous wicks (30 kN) > porous wicks (40 kN) > porous wicks (50 kN) > porous wicks (60 kN), which is consistent with the porosity test results.

3. Structure design of loop heat pipe

3.1 Structural design of capillary cores

The structure size of capillary core has an important effect on the performance of loop heat pipe. For example, the length of capillary core has a significant effect on the heat transfer capacity of loop heat pipe. When the length of the capillary core is long, the synergy between the temperature and the flow field becomes longer with the increase of the length of the capillary core, but when the capillary core is too long, the flow resistance of the working fluid in the capillary core is further increased. It will result in less fluid flowing through the capillary core in unit time and affect the heat transfer rate of the loop heat pipe. The structural size of the capillary core was prepared with reference and in the second chapter of this paper. The final size of the capillary core was 100 mm in length and 20 mm in diameter. The 3D model of capillary wick is as shown in **Figure 13.**

The capillary core is not only a simple cylindrical structure, its structure is relatively complex, the outside is a vapor trough for the passage of vapor, and the internal storage tank for liquid working fluid. The curved liquid surface of vapor liquid

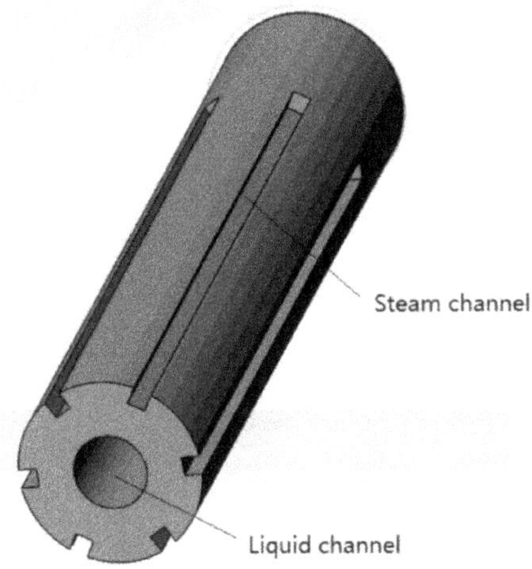

Figure 13.
Schematic diagram of capillary core structure.

phase transition of the working fluid is in the capillary core, the liquid working fluid enters through the liquid channel, and the vaporization at the curved liquid level is derived from the vapor tank. The results show that the ratio of the depth to width of the capillary vapor channel is 1:1, and the effect is the best when the length of the groove is 75 mm, the depth and width are 2 mm, and the number of grooves is 6. In the third chapter, the capillary core suction ability of different inner diameter liquid storage channel is studied. The capillary core suction performance is the best when the inner diameter is 8 mm. The final structure of the evaporator and reservoir is shown in **Figure 15**.

3.2 Design of evaporator and liquid tank

Evaporator and liquid accumulator are the main components of loop heat pipe. Especially the evaporator, which contains capillary core, is the place where the liquid working fluid changes to the gas working medium in the loop heat pipe. However, the vapor generated in the capillary core must prevent its reverse flow into the liquid reservoir. In the evaporator, there is a liquid lead pipe in the capillary core, which is connected to the vapor pipeline, and the liquefied working fluid is directly introduced into the liquid channel inside the capillary core. The structure of the evaporator is shown in **Figure 14** below. Structural diagram of evaporator and liquid reservoir is shown in **Figure 15**.

In the design of loop heat pipe evaporator and liquid accumulator, the most important thing is to ensure the positive heat conduction of the loop heat pipe, and the most important thing is to do the sealing work well. The seal of evaporator and liquid accumulator means that there is a certain pressure bearing capacity in isolation from the outside environment, and more important is to prevent the diffusion (heat leakage) of the gas working fluid from the evaporator to the liquid accumulator. The heat transfer within the loop heat pipe is not strictly unidirectional, but the heat transfer in the loop heat pipe is not strictly unidirectional. Most of the external heat input in the capillary core makes the working fluid gasification to participate in the loop heat pipe circulation, only a part of the heat into the liquid reservoir in the form of heat conduction, this part of the heat called is a heat leak. Heat leakage will lead to excessive temperature and abnormal increase of pressure of the liquid accumulator, which will affect the normal operation of the working fluid in the loop heat pipe.

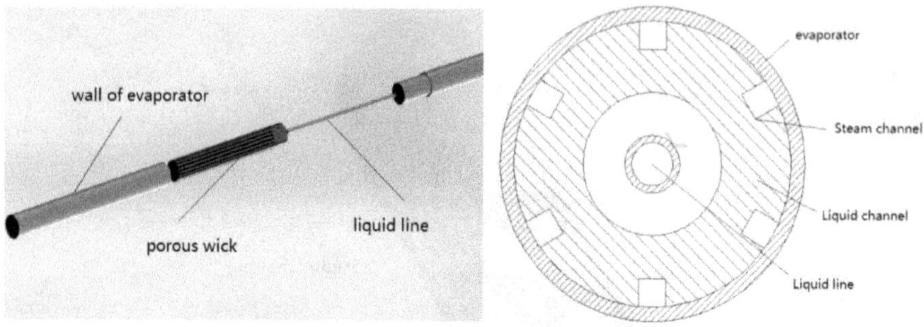

Figure 14.
Evaporator structure profile.

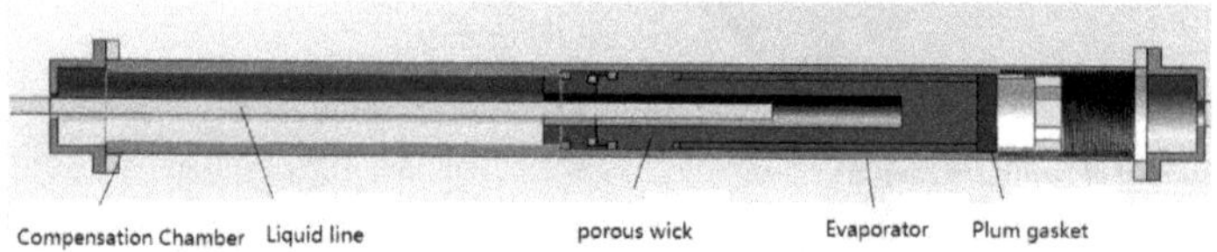

Figure 15.
Structural diagram of evaporator and liquid reservoir.

The heat leakage is mainly transmitted through the heat conductivity of the capillary core and the outer wall of the loop heat pipe, which is difficult to avoid. However, the heat leakage should be considered in the preparation of the loop heat pipe. Limit the adverse effects of heat leakage to a lower range of effects. If vapor enters the tank, it will cause the temperature and pressure of the tank to rise, which will lead to the failure of the heat pipe operation. When the vapor is running in the opposite direction, the heat transfer efficiency of the loop heat pipe will be seriously affected, which will cause the gas accumulation in the liquid accumulator and the capillary seriously. The thin core leads to the failure of forward heat and mass transfer in the loop heat pipe.

In order to prevent the reverse operation of vapor, an inner step stainless steel outer wall is designed for evaporator and liquid accumulator in order to reduce welding, and the capillary core is clamped in one direction, and a clasp structure is installed at the capillary core and step. There are three gaskets on the outer wall of the evaporator and liquid accumulator. Layer by layer protection reduces the reverse flow of heat vapor along the inner wall, improves the heat transfer energy of the return heat pipe, and simulates the failure of the heat pipe operation. Evaporator wall thickness as thin as possible to reduce thermal resistance to facilitate the capillary core to absorb external heat. At the vapor outlet, the capillary core is clamped by thread connection, and a plum flower gasket is installed between the bolt structure and the capillary core, and the vapor produced by the thermal reaction flows out of the pore between the pores of the plum flower gasket. This junction can be reused, just loosen the bolt structure to replace the capillary core. The specific size is obtained from the size of the capillary core. As shown in **Figure 15**.

3.3 Condenser optimization

In the loop heat pipe system, the condenser is responsible for the rapid transfer of heat from the evaporator to the outside world. After heated vaporization of the working medium in the evaporator, the hot vapor enters the condenser through the gas pipeline, and the heat is exchanged with the outside in the condenser to dissipate heat towards the outside world, and the vapor moves towards the liquid storage device after the condenser becomes a liquid. Therefore, the condenser must have sufficient undercooling to ensure that the working fluid can be completely condensed into liquid. At present, the cooling methods of various equipment are water cooling or air cooling. It is found that the vapor in the evaporator is usually difficult to be condensed into liquid due to the lack of cooling power. So there are a lot of heat pipes in the loop. Adopt water cooling. In order to further improve the cooling temperature, alcohol is chosen as the cooling medium.

The loop heat pipe condenser prepared in this paper is shown in **Figure 16**. The condenser adopts a cylindrical tube structure with an inlet and an outlet to circulate low temperature alcohol or cold water, and the cylinder pipe is a cooling pipe. Considering the overall size of the loop heat pipe, the length of the condenser is set to 200 mm. However, the heat transfer length of 200 mm is insufficient and the vapor in the condenser is cooled completely into a liquid. Faced with this situation, there are usually two solutions: one is to put fins on the outside of the pipe, the other is to make the pipe into a coil. After the experiment, it was found that the efficiency of the first scheme was also insufficient when the first scheme was running at high power. Therefore, the coiled tube alcohol cooling was used in the end. However, as the final cooling scheme, we should pay attention to the length of coil should not be too long, too long pipe length will lead to excessive resistance in operation of the working fluid, which will affect the forward operation of the loop heat pipe.

3.4 Loop heat pipe assembly

The connection between evaporator and condenser in heat pipe is vapor pipeline and liquid pipeline. In order to reduce the flow resistance of the medium, the more smooth the inner, the better. In this paper, stainless steel tube is selected and the inner polishing is done. The vapor liquid phase change process is involved in the operation of the loop heat pipe, and the sealing property is very high. Therefore, stainless steel is used in all parts of the loop heat pipe, and the connection between each part is argon arc welding. All stainless steel components (including liquid accumulators and evaporators) should be cleaned according to the stainless steel cleaning method before final use to improve the performance of the loop heat pipe. The loop heat pipe also needs to be equipped with a new belt valve. The working fluid filling mouth of the door is filled and sealed with heat pipe working fluid. After welding the loop heat pipe, it is necessary to pick up the leakage of the system, inflate the system with air compressor from the filling port, and place the system

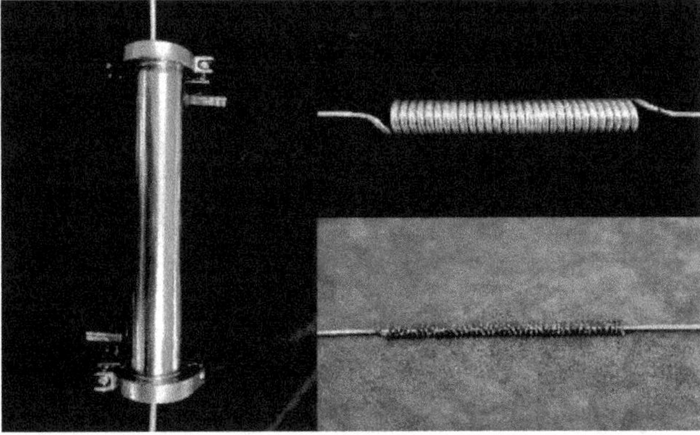

Figure 16.
Condenser structure (including two heat dissipation modes).

in water. If there is no bubble, the system is sealed completely, and if there is air bubble, the leakage point of welding should be rewelded. In order to reduce the difficulty of welding and to facilitate the performance experiment in the future, the loop heat pipe is made into a rectangle, 1000 mm in length and 300 mm in width, in order to match the heat transfer test bed designed later.

The structure and dimensions of the heat pipe are described in **Figure 17** and **Table 1**.

3.5 Vacuum and perfusion of loop heat pipes

In addition to the capillary core properties, the internal working fluid perfusion of the loop heat pipe has the greatest influence on the performance of the loop heat pipe. There are two factors that influence the perfusion effect: one is the quality of perfusion, the other is the degree of vacuum during perfusion, that is, the purity of the injected working fluid. First of all, the heat pipe is filled with flux, when the charge is too small, it will cause the heat pipe to be burned out in the liquid storage device of the heat pipe evaporator, and the heat pipe will fail. When the charge is too much, there will be a lot of liquid working fluid in the condenser, which will cause excessive resistance along the path and hinder the forward operation of the heat pipe. It is inevitable to mix air and other non-condensable gases when pouring working fluid into the loop heat pipe. The research of Beihang [84DIAN85] finds Danghuan when there are non-condensable gases such as nitrogen in the heat pipe, it will lead to the difficulty of starting the loop heat pipe, the high temperature of the evaporator and

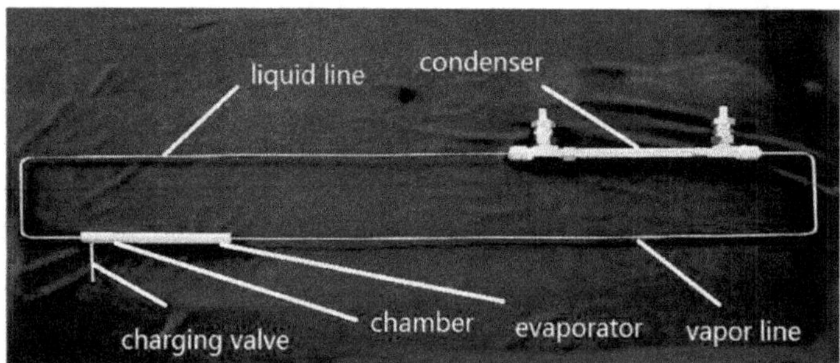

Figure 17.
Loop heat pipe structure diagram.

Loop heat pipe component	Dimension parameter(mm)
Evaporator (Dout/Din/L)	26/20/150
Reservoir (Dout/Din/L)	26/18/120
Capillary core (D/L)	20/100
Capillary vapor groove (L/W/H)	80/2/2
Capillary core liquid channel (Din/L)	8/80
Vapor line (Dout/Din/L)	6.35/3.89/1200
Liquid line (Dout/Din/L)	6.35/3.89/1200
Condenser pipeline (Dout/Din/L)	6.35/3.89/2000
Loop heat pipe appearance (L/W)	1000/300

Table 1.
Size diagram of loop heat pipe.

the decrease of the heat transfer efficiency. It is also found that the non-condensable gas will greatly reduce the service life of the loop heat pipe. Therefore, it is necessary to maintain a vacuum environment during perfusion. We should pay attention to the leakage of the loop heat pipe and the leakage of the working fluid during the simulation of the perfusion according to the way mentioned in this paper.

3.6 Vacuum infusion

Loop heat pipe perfusion is divided into two important steps: first, vacuum, and secondly, perfusion. Therefore, a loop heat pipe perfusion platform integrating vacuum pumping and perfusion is established in this paper. Based on dual functions, the pipeline must have two routes. The first is the perfusion pipeline, which is used to introduce liquid ammonia into the loop heat pipe. The perfusion line is shown in **Figure 18** above.

The pipeline is designed from top to bottom, the top is a liquid ammonia bottle, and the bottom is connected with a loop heat pipe. The loop heat pipe outlet is a pressure reducing valve to monitor the liquid ammonia outlet and the pressure in the pipeline. Then there are two condensing units to try to prevent liquid ammonia from liquefaction. Ensure the accuracy of subsequent flowmeter measurements and prevent too much gas in the pipeline from fluctuating the flow rate. The flowmeter is used to detect the amount of heat in the heat pipe. Continue after a one-way valve to prevent working fluid backflow. One-way valve is a flow valve. The pressure reducing valve can control the flow velocity of the working fluid in the pipeline to control the filling speed. The flow valve is followed by the valve and the loop heat pipe.

The exhaust line also consists of two parts, one is the air in the perfusion line and the other is the air in the heat pipe of the loop, as shown in **Figure 19**. The design scheme of the main line of the heat pipe pumping vacuum perfusion test rig is shown in **Figure 22**. The valve which combines the pouring pipe and the vacuum pumping pipe to control the pipe passage condition is designed as shown in **Figure 20**.

The designed pipe is placed vertically, the aluminum alloy frame is arranged, and the flowmeter display device is designed to monitor the quality of the working fluid that has been poured in the heat pipe. During perfusion, the heat pipe accumulator should be placed in a lower temperature environment than the perfusion tube condenser. This is because it is difficult for liquid ammonia to flow into a loop heat pipe simply because of the gravity effect, so that the temperature in the heat pipe is lowered so that the pressure inside the loop heat pipe remains relatively low all the time. The flow of liquid ammonia in the pipeline is promoted by the pressure of different positions in the pipeline. The vacuum pumping system designed

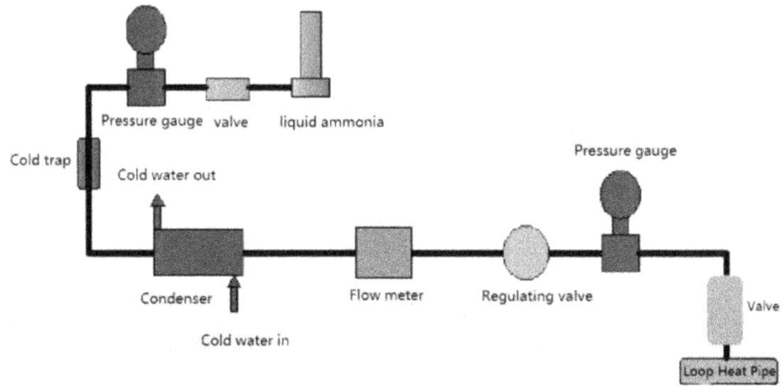

Figure 18.
Perfusion pipeline diagram.

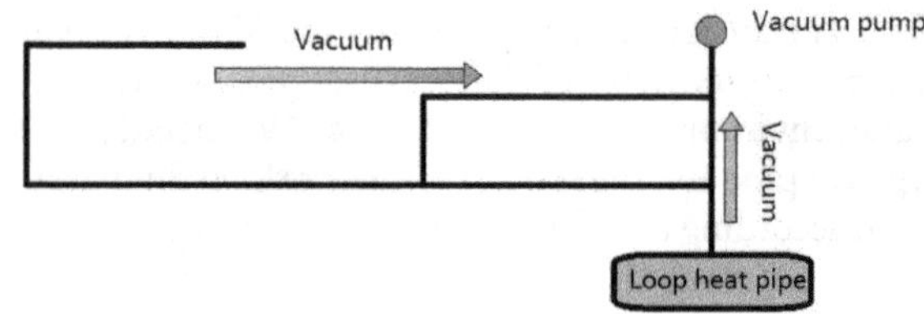

Figure 19.
Drawing of vacuum line.

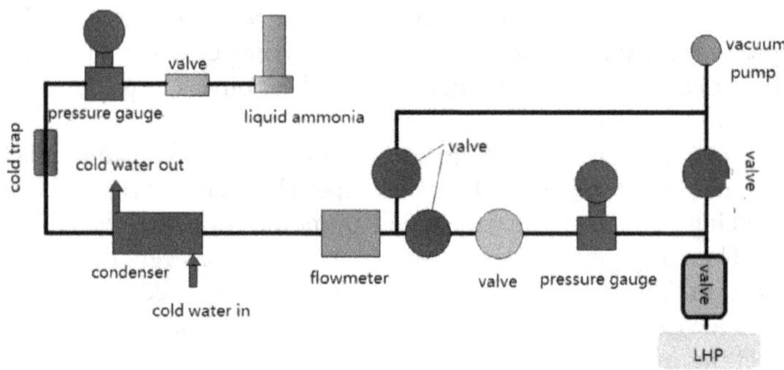

Figure 20.
Schematic diagram of vacuum perfusion pipeline.

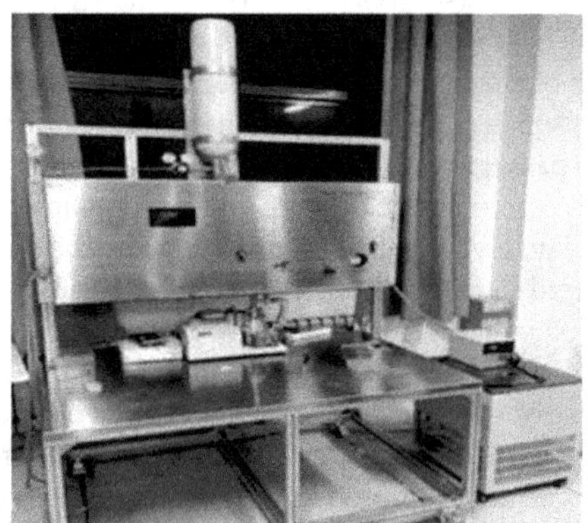

Figure 21.
Loop heat pipe vacuum perfusion test bench.

in this paper uses Edward molecular pump to pump the vacuum in the pipeline. In the vacuum process, the pressure in the pipeline can be lowered by 6×10^{-2} Pa. The device diagram is shown in **Figure 21.**

4. Heat transfer experiment of loop heat pipe

In this paper, the heat transfer test bench of loop heat pipe is built, including heating module, cooling module, temperature collecting module, etc. The performance of loop heat pipe is studied experimentally. The loop heat pipe test platform consists of three modules: heating module, cooling module and temperature collecting module. Heating module has two sets, one is controlled power heating, the other is constant temperature heating. The output power is controlled by puss power supply, and the loop heat pipe evaporator is heated by resistance wire. Constant temperature heating using cast copper heating block. There are grooves on the surface of cast copper heating block for heating loop heat pipe evaporator. The cast copper heating block is controlled by PID and the temperature is monitored by internal thermocouple. This heating method can only guarantee the outer wall temperature of the loop heat pipe evaporator without knowing the heating power of the loop heat pipe. According to the situation, these two heating methods should be used.

The experimental platform needs a cooling module to condensate the loop heat pipe condenser. The condenser is connected to the DC20-20 type low temperature constant temperature tank, and the flow velocity of the constant temperature tank is 20 L/min. The lowest temperature can be −20°C, so this tank can be filled with alcohol as a cooling agent.

The data acquisition module is used to collect the local temperature of the heat transfer experiment of the loop heat pipe, and the performance of the loop heat pipe is analyzed by the variation of each temperature. The data collector uses Fluke2638A to collect data, and Fluke2638A has three chucks, each of which has 20 channels, which can collect 60 temperature points at the same time. K-type Ω thermocouple is used to collect temperature transmission data to data collector. The thermocouple is made of TT-K-36 type Ω temperature measuring line, and MES thermocouple welding machine is used to weld the temperature measurement line. The temperature range of the thermocouple was −200–260°C, and the measurement error was ±0.5°C. In order to ensure the detection accuracy, each K thermocouple is connected to the data collector channel and the ice water is mixed to zero.

The distribution of thermocouple is shown in **Figure 22**, (1) is the outlet temperature of liquid pipeline, (2) is the inlet temperature of the liquid reservoir, (3) is the inlet temperature of the evaporator, (4) is the temperature at the heating point, (5) is the inlet temperature of the vapor pipe, (6) is the intermediate temperature of the vapor pipe, (7) and (8) are the temperature at the inlet and outlet of the condenser, and (9) are the temperature in the middle of the liquid pipeline. In addition, the condenser inlet and outlet temperatures need to be measured.

4.1 Study on start-up characteristics of loop heat pipe

Eva 1 is the temperature of evaporator, ***eva 2*** is another temperature of evaporator, ***s line 1*** is the temperature of vapor line near the evaporator, ***s line 2*** is the temperature of vapor line near the condenser, ***conin*** is the temperature of the condenser inlet, ***conout*** is the temperature of the condenser outlet, ***waterin*** is the temperature

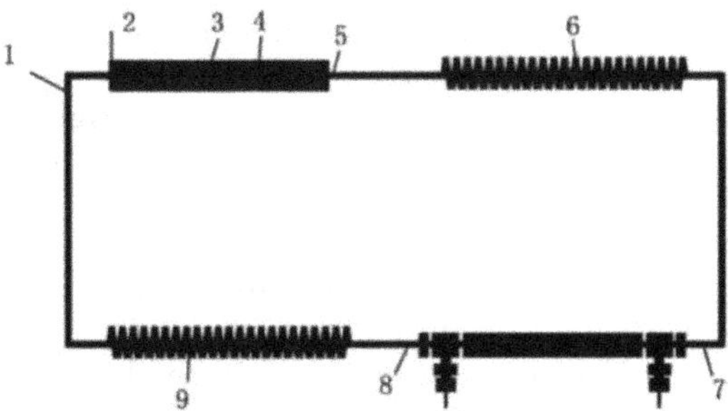

Figure 22.
Location of thermocouple distribution.

of water of the condenser inlet, ***waterout*** is the temperature of water of the condenser outlet, ***l line 1*** is the temperature of the liquid line near condenser, ***l line 2*** is the temperature of the liquid line near chamber, ***CC 1*** is the temperature of chamber, ***CC 2*** is another temperature of chamber, ***steamout*** is the temperature of vapor in the outlet of evaporator.

The startup of LHP can be divided into the following four processes: (1) after the evaporator is heated, the heat is transferred to the working fluid and vaporized. Because of the barrier of the capillary core sintering structure, the vapor can only enter into the vapor pipe through the vapor trough; (2) the vapor enters the condenser to cool through the vapor pipe, (3) after condensing into the liquid in the condenser, the liquid working fluid is pumped back into the liquid storage room because of the capillary force of the capillary core; (4) the working fluid of the reflux flows back into the evaporator through the permeability of the capillary core. In this paper, it is found that the minimum starting power of the loop heat pipe is 5 W.

Figure 23 shows the starting heat transfer characteristics of the loop heat pipe at 5 W startup. It can be seen from the diagram that the loop heat pipe operates steadily after a period of time (3500 s), and the temperature of each part remains constant. In the early stage of the experiment, the evaporator and vapor line begin to rise steadily, the internal working fluid of the loop heat pipe evaporator absorbs the external heat, the internal working fluid reaches the saturation temperature and begins to vaporize, and the vapor enters the vapor pipeline, which leads the internal temperature of the pipeline to increase. Vapor passes through the vapor line and enters the condenser, so the inlet temperature of the condenser jumps. Through the condenser enough gas refrigerants are condensed into the liquid line, so the liquid tube is filled with liquid. The line temperature is low. After that, the liquid working fluid is reflowing back to the liquid accumulator and evaporator under the action of the capillary core to complete the forward circulation of the loop heat pipe. At the initial stage of starting the loop heat pipe, there was a small fluctuation in the temperature at the evaporator. The temperature first decreased and then increased. This was caused by the liquid working fluid began to condensate and reflux, and it also marked the formal start of the loop heat pipe. The starting time of the loop heat pipe is 1000 s.

4.2 Study on heat transfer characteristics of loop heat pipe

The threshold value for the initial start-up of the loop heat pipe is 5 W. When the heating power of the loop heat pipe evaporator is increased, the temperature of each part of the loop heat pipe is different. In order to further understand the heat transfer performance of the loop heat pipe prepared in this paper, the heating power of the loop evaporator is gradually increased, and the relationship between the temperature of each part of the loop heat pipe and the heating power is discussed. In this paper, the heat transfer experiment of loop heat pipe with different heating power has been carried out from 5 to 1 W interval. The experimental scheme is carried out according to the above experimental scheme. The external cold source temperature is −5°C when the heat transfer is heated by the transverse power heating method. The temperature curve of heating power 8W and 10W with −5°C heat sink is as shown in **Figures 24** and **25**.

With the increase of evaporator power, the loop heat pipe can continue to operate normally, and its operation law is similar to that of each part of the heat pipe heated at 5 W. The temperature inside the evaporator and vapor line rises first and then remains stable. The inlet temperature of the condenser is also stable after rising sharply at the beginning, and the temperature difference between the inlet of the condenser and the inlet of the condenser has been maintained, indicating that the loop heat pipe has been in a positive operating state. The experimental results show that the liquid line temperature of the loop heat pipe is approximately the same under different heating power, which indicates that the working fluid can be condensed completely into liquid through the condenser when the loop heat pipe is running, and is lower than the saturation temperature under the current pressure. The temperature curve and thermal resistance of different heating power are shown in **Figures 26** and **27**.

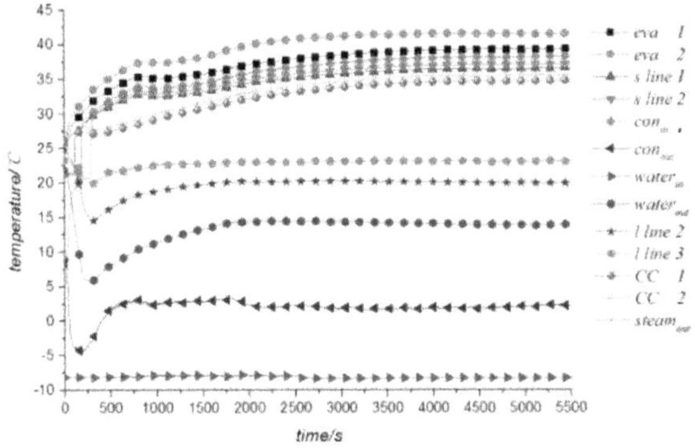

Figure 23.
5 W heating power loop heat pipe temperature curve.

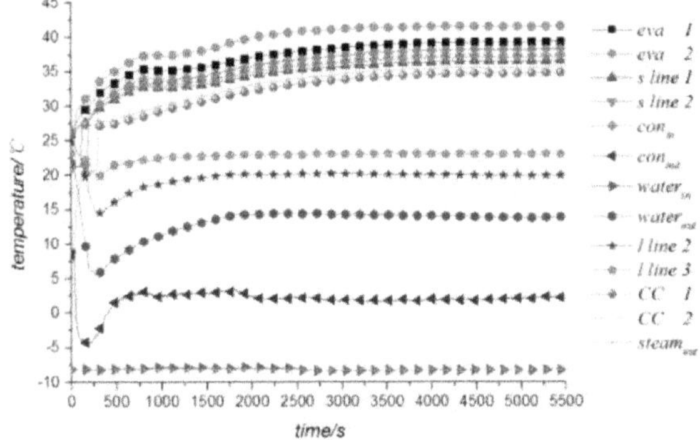

Figure 24.
8 W heating power loop heat pipe temperature curve (−5°C heat sink).

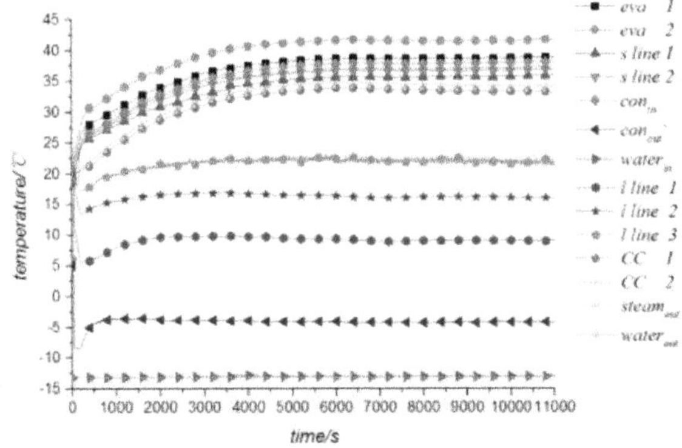

Figure 25.
10 W heating power loop heat pipe temperature curve (−5°C heat sink).

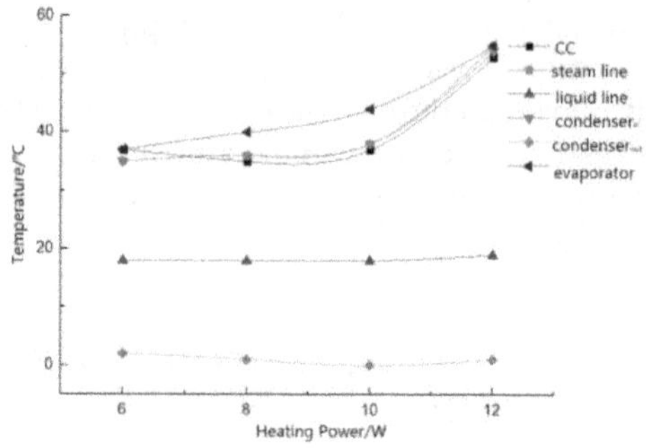

Figure 26.
Curve of temperature change with heating power.

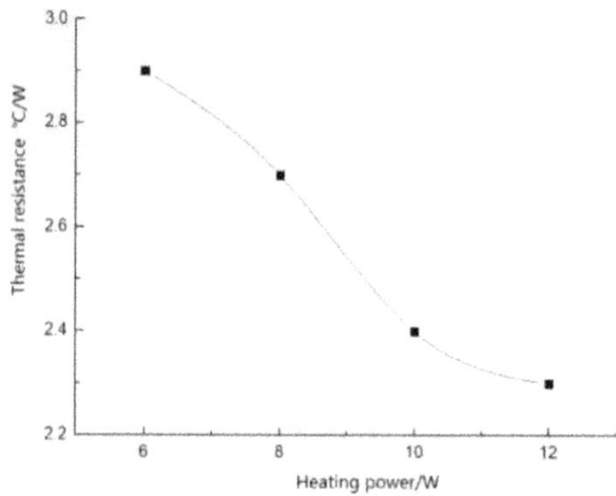

Figure 27.
Relationship between heating power and thermal resistance.

The temperature variation of loop heat pipe is different with different heating power. As is shown that the higher the heating power, the higher the temperature of the evaporator, the higher the internal pressure, and the faster the temperature of the evaporator increases with the increase of heating power, so the heating power of the loop heat pipe should not be too large. Prevent the internal pressure from exceeding the pressure limit of the loop heat pipe. With the increase of the evaporator temperature, the temperature of the vapor pipeline entering the gaseous medium also becomes higher, which leads to the continuous increase of the vapor pipeline temperature to the inlet temperature of the condenser. In this paper, the low temperature alcohol is used as the external heat sink in the experiment, and the loop heat pipe is condensed. The working fluid in the condenser can be condensed completely, so even though the external heating conditions are different, the outlet temperature of the condenser remains basically unchanged at about 0°C. The experimental results show that the temperature of the loop heat pipe liquid accumulator increases rapidly with the increase of heating power, which is due to the fact that the evaporator tube wall and the liquid storage tube wall are made from the same stainless steel jacket. The heat from the outer wall of the evaporator is transferred to the outer wall of the liquid storage device by heat conduction, which results in the increase of the temperature of the external wall of the liquid storage device. The increase of external wall temperature may lead to excessive internal temperature and high pressure, which may hinder the forward operation of the loop heat pipe. So in subsequent studies, How to reduce the heat leakage from evaporator to liquid accumulator will be an important and difficult problem.

The heat transfer efficiency of loop heat pipe is usually measured by its overall thermal resistance. The total thermal resistance (R_{total}) of the loop is expressed as follows: the difference between the average temperature of the evaporator and the average temperature of the condenser is used to compare the heating power of the upper loop heat pipe with the difference between the average temperature of the evaporator and the average temperature of the condenser.

$$R_{total} = \frac{T_{ev} - T_{cool}}{Q_{load}} = \frac{T_{ev}^{in} + T_{ev}^{out} - T_{cool}^{in} - T_{cool}^{out}}{2Q_{load}} \tag{4}$$

The experimental results show that the thermal resistance of the loop heat pipe becomes smaller and the overall heat transfer performance of the loop heat pipe becomes more and more excellent when the heating power is increasing, that is to say, the thermal conductivity of the loop heat pipe is getting better and better. However, the heat resistance of the loop heat pipe has a minimum value. The heat transfer efficiency of the loop heat pipe is the highest when the loop heat pipe works under this condition, but at the same time the temperature inside the evaporator is also very high, and the heat transfer limit of the loop heat pipe reaches its heat transfer limit.

The experimental results show that when the external condensation temperature is insufficient, the heat absorbed by the loop heat pipe evaporator cannot be transferred out, which leads to the high pressure inside the loop heat pipe. The capillary force provided by the capillary core cannot overcome the resistance in the loop heat pipe normally and the suction of the working fluid leads to the failure of the loop heat pipe operation. However, when the condenser and the external heat exchange is sufficient, the evaporator can transfer the heat completely, and the different undercooling degree has no great influence on the steady state of the loop heat pipe. As is shown in curves of heat sink temperature of −10°C at −15°C under heating power of 10 W of the loop heat pipe. It can be seen that the temperature distribution of loop heat pipe evaporator, liquid accumulator, vapor line and liquid line are basically the same at different external heat sink temperatures. The temperature at the outlet of the condenser is affected by the external heat sink of the condenser. The lower the external heat sink temperature, the lower the temperature of the condenser, and the difference of the temperature at the outlet of the condenser is exactly the difference of the external heat sink. The external heat sink temperature also affects the time required for the loop heat pipe to reach stable operation. It can be seen from the diagram that the lower the external heat sink

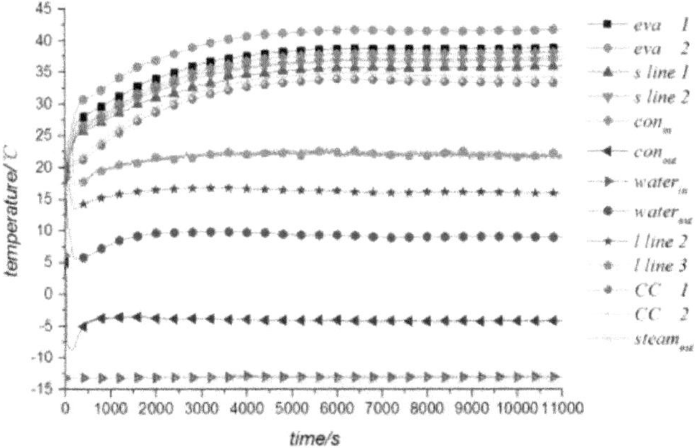

Figure 28.
Temperature curve of heating power loop heat pipe (−10°C heat sink).

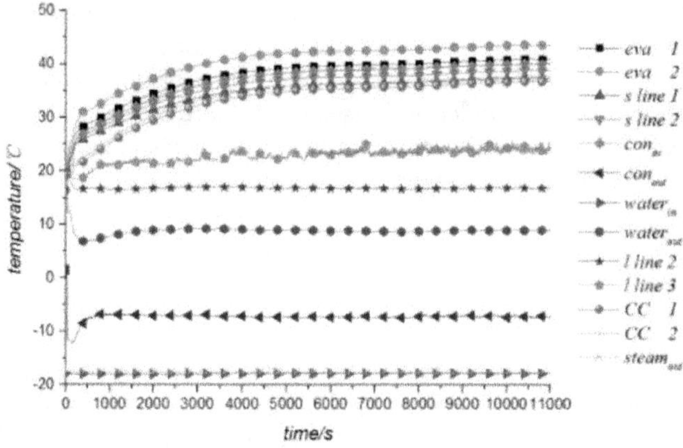

Figure 29.
Temperature curve of heating power loop heat pipe (−15°C heat sink).

temperature, the longer the loop heat pipe operation will be stable. The temperature curve of different heat sink (–10°C and –15°C) are shown in **Figures 28** and **29**.

5. Conclusion

1. With the increase of mass of NaCl, the suction speed and total suction mass increase, it can be concluded that the pore-forming agent (NaCl) will significantly improve capillary core performance, which is crucial for the loop heat pipe.

2. In select range, the cold forming pressure has little influence on the performance of the capillary cores.

3. The loop heat pipe which uses the novel capillary cores can start up successfully on 5 W. Besides the heat sink will influence the time that the LHP operate stably. The lower the temperature of the external heat sink, the longer the loop heat pipe will run stably.

Author details

Chunsheng Guo
School of Mechanical, Electrical and Information Engineering, Shandong University, Weihai, China

*Address all correspondence to: guo@sdu.edu.cn

References

[1] Maydanik YF, Chernyshevana MA, Pastukhov VG. Review: Loop heat pipes with flat evaporators. Applied Thermal Engineering. 2014;**67**:294-307

[2] Maydanik YF. Loop heat pipe. Applied Thermal Engineering. 2005;**25**:635-657

[3] Huang BJ, Hsu PC. A thermoelectric generator using loop heat pipe and design match for maximum-power generation. Applied Thermal Engineering. 2015;**91**:1082-1091

[4] Baumann J, Cullimore B. Non-condensable gas, mass, and adverse tilt effects on the start-up of loop heat pipes. The International Journal of Automotive Engineering. 1999;**1**:2036-2048

[5] Zhang HX, Lin GP, Ding Z, Shao XG, Sudakov RG, Maydanik YF. The experimental study of loop heat pipe start-up characteristics. Journal of Science China's Scientific E Series: Engineering Science. 2005;**35**:1720-1730

[6] Yan T, Zhao YN, Liang J, Liu F. Investigation on optimal working fluid inventory of a cryogenic loop heat pipe. International Journal of Heat and Mass Transfer. 2013;**66**:334-337

[7] Mitomi M, Nagano H. Long-distance loop heat pipe for effective utilization of energy. International Journal of Heat and Mass Transfer. 2014;**77**:777-784

[8] Tang Y, Deng D, Huang G, Wan Z, Lu L. Effect of fabrication parameters on capillary performance of composite wicks for two-phase heat transfer devices. Energy Conversion and Management. 2013;**66**:66-76

[9] Bai L, Lin G, Zhang H, Miao J, Wen D. Operating characteristics of a miniaturecryogenic loop heat pipe. International Journal of Heat and Mass Transfer. 2012;**55**:8093-8099

[10] Du C, Bai L, Lin G, Zhang H, Miao J, Wen D. Determination of charged pressureof working fluid and its effect on the operation of a miniature CLHP. International Journal of Heat and Mass Transfer. 2013;**63**:454-462

[11] Ku J, Ottenstein L, Douglas D, et al. Multi-Evaporator Miniature Loop Heat Pipe for Small Spacecraft Thermal Control, Part 1: New Technologies and Validation Approach[C]. In: 48th AIAA Aerospace Sciences Meeting Including the New Horizons Forum and Aerospace Exposition Orlando, Florida. 2010;**6**:2010-1493

Chapter 4

Study of a Novel Liquid-Vapour Separator-Incorporated Gravitational Loop Heat Pipe

Xudong Zhao, Chuangbin Weng, Xingxing Zhang, Zhangyuan Wang and Xinru Wang

Abstract

The aim of this chapter is to report the study of a novel liquid-vapour separator-incorporated gravity-assisted loop heat pipe (GALHP). This involves a dedicated conceptual formation, thermo-fluid analyses, and computer modelling and experimental validation. The innovative feature of the new GALHP is the integration of a dedicated liquid vapour separator on top of the evaporator section, eliminating the potential entrainment between the heat pipe liquid and the steam stream, while addressing the inherent 'dry-out' problem exhibited in the traditional GALHP. Based on this recognised novelty, a dedicated steady-state thermal model covering the mass continuity, energy conservation and Darcy equations were established. Under the specifically defined operational condition, the proposed GALHP has more evenly distributed axial temperature profile. The effective thermal conductivity in the proposed GALHP was 29,968 W/C m. It is therefore concluded that the novel heat pipe could achieve a significantly enhanced heat transport effect. The results derived from this research enabled characterisation of the thermal performance of the proposed GALHP and validation of the developed computer simulation model. The research will enable design, optimisation and analysis of such a new GALHP, thus promoting its wide application and achieving efficient thermal management.

Keywords: LHP, composite wick, start up, thermal conductivity

1. Background

A heat pipe [1, 2] is an effective heat transfer device functioning through the evaporation and condensation cycles without external driving forces, which is normally used under three circumstances: (1) transferring heat from the heat source to heat sink at a distance; (2) transporting a thermal shunt in an effective way and (3) dissipating heat effectively across a plane. To date, various types of heat pipes have been investigated for various applications, for example, electronics cooling, solar devices, heat exchangers, aerospace, medical applications and transportation system [3]. Of these, the most commonly used one is a gravitational straight type, which possesses the problems of increased flow resistances within the vapour and liquid flows and reduced overall heat transport capacity of the heat pipe.

Loop heat pipe (LHP) [3–5] is a two-phase (liquid/vapour) heat transfer device allowing a high thermal flux to be transported over a distance of up to several tens of metres in a horizontal or vertical position owing to its capillary or gravitational structure. LHP has a separate evaporator and condenser, thus eliminating an entrainment effect occurring in between. LHP can operate under different gravitational regimes, regardless of whether the evaporator is above or below the condenser.

A conventional LHP is usually composed of the complex capillary pumps (evaporators), compensation chambers (storage), condensers and vapour and liquid transfer lines [6–8]. The working principle of the LHP device could be described as follows [1]: the heat transfer fluid in the wick absorbs the heat added to the evaporator and vaporises via the vapour line to the condenser. Within the condenser, the vapour will be condensed to the liquid of the same temperature and return to the compensation chamber through the liquid line. The liquid will then be accumulated and stored in the compensation chamber and further saturate the wick.

Numerous works in relation to LHP have been developed, for example, loop component designs, mathematical models, working fluid and wick structures. The first LHP was developed and tested in 1972 by Russian scientists Gerasimov and Maydanik [9]. A book written by Peterson [10] illustrated the performance limit approach for the heat pipe in the steady-state condition. Peterson [11] also analysed the heat pipe's heat transfer processes in the steady-state condition by using thermal resistances calculating method. In order to simplify the existing engineering models and reduce the required computing resources, Zuo and Faghri [12] developed a thermal network model to analyse the circulation of the working fluid in the heat pipe by using the thermodynamic cycle approach. Kaya and Hoang [13] modelled the performance of a LHP based on steady-state energy balance equations at each component of the loop. The loop operating temperature was found to be a function of the applied power at the given loop condition. Bai et al. [14] established a mathematical model for the start-up process of a LHP based on the node network method. Pauken and Rodriguez [15] modelled and tested a LHP with two different working fluids, that is, ammonia and propylene. Hoang et al. [16] mentioned that the heat transfer characteristic of a LHP was difficult to predict, owing to the complicated nature of the thermal interaction between the LHP and environment. Riehl [17] tested a LHP system operating with acetone as the working fluid. Zan et al. [18] established an experimental formula for a sintered nickel powder wick. Riehl and Dutra [19] presented the development of an experimental LHP. Vlassov and Riehl [20] explored LHP modelling by developing a relatively precise condenser sub-model from the solutions of the conjugate equations of energy, momentum and mass balances, and only describing a few transient nodes within the evaporator and compensation chamber. A more comprehensive dynamic model was published by Launay et al. [21], who proposed a transient model to predict the thermal and hydrodynamic behaviour of a standard LHP.

In recent years, the application of the LHP in solar thermal field has become more and more attractive owing to the significant technical advance in renewable energy [22–24]. LHP is suitable for use in building solar hot water system, owing to its unique features, that is, highly effective thermal conductance and flexible design embodiment and installation [1]. For such an application, the LHP was mostly operated under gravity-assisted conditions, and termed gravitation-assisted loop heat pipe 'GALHP'. The GALHPs have been identified with two shortfalls that need to be tackled with, that is, complicated wick structure and liquid film 'dry-out' problem [3, 4].

In order to overcome the problems exhibited by the conventional GALHP, a novel liquid-vapour separator-incorporated GALHP was proposed, which is dedicated to simplify the wick structure, eliminate the 'dry-out' potential and, thus, create a high-efficient and cost-effective heat transport solution. Through the theoretical and experimental analysis, the analysis results will be compared with the conventional GALHP and conventional straight heat pipe. The research results could be directly used for design, optimisation and analyses of the new GALHP configuration, thus promoting its wide applications in various situations to enable the enhanced performance of the GALHP heat transport to be achieved.

2. Description of the proposed GALHP

Schematic of the proposed GALHP is shown in **Figure 1**, and the novel liquid-vapour separator-incorporated GALHP is shown in **Figure 2**. This separator is configured as a three-way structure, internally containing a tubular pipe with a downward expanding opening, which is fitted into the top of the evaporator while the edge of the expanding opening is tightly attached to the wicked inner surface of the heat pipe. In this way, the return liquid will be reserved in the liquid reservoir above the evaporator, thus formulating a certain liquid head. Under the action of the liquid head, the liquid will penetrate through the peripheral gap between the pipe and the expanding opening, and flow evenly downward along the wicked surface of the heat pipe. Meanwhile, the evaporated fluid, in the form of vapour,

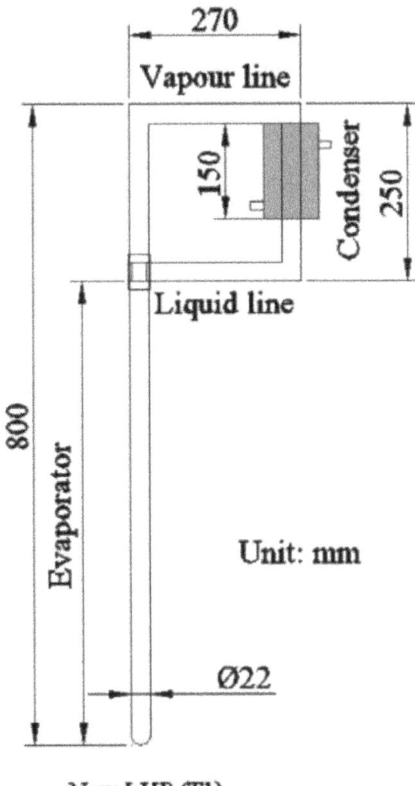

Figure 1.
Schematic of the proposed GALHP.

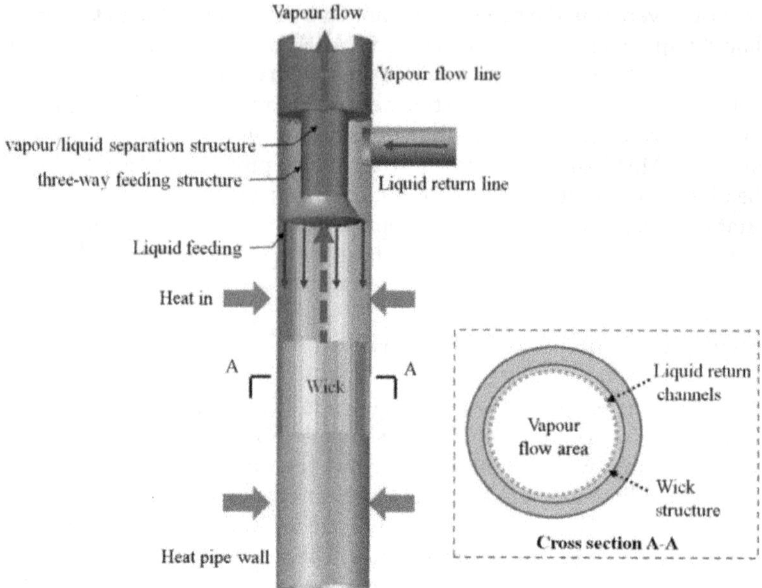

Figure 2.
Top-positioned vapour-liquid separator of new GALHP.

will flow upward through the central tubular pipe and enter into the vapour transport line. If the liquid level is further controlled by a valve mounted on the liquid transfer line, the downward liquid flow rate can be controlled to match the rate of evaporation on the inner surface of the heat pipe. So the wicked inner heat pipe surface will be constantly in 'wet' state, thus preventing the potential 'dry-out' problem with the conventional GALHP. Meanwhile, the vapour and liquid flows will be regulated in the same direction and separated clearly during the operation, thus preventing the potential entrainment problem with the conventional straight heat pipe.

Based on the above innovative concept, a dedicated mathematical model and the associated computer program will be developed to analyse the characteristics of the new GALHP.

3. The model applicable to the proposed GALHP

The mathematical model for the evaporation-condensation processes in the proposed GALHP is based on the following assumptions:

- Heat transfer and fluid flow occur under the quasi-steady-state condition.
- Heat conduction and fluid flow across the wick are one-dimensional in the radial direction.
- Heat pipe evaporator is heated axial-symmetrically and the difference of the temperature along the axial direction is negligible.
- The hydrostatic pressure drop across the radial direction owing to the gravity effect is considered to be zero.

- The axial pressure drop is negligible due to less magnitude against the gravitational head.
- The working fluid is incompressible and has a constant property value in each phase.
- The wick is liquid saturated and wick material is assumed homogenous and isotropic.
- A local thermal equilibrium exists between the porous structure and the working fluid.
- Heat loss to the surroundings is ignored due to the well-insulated pipes.

The mass flow rate within the wick structure is considered to be constant owing to the mass conservation law, given by [4, 16]

$$\dot{m} = \rho_l \mu_l 2\pi \gamma L_{eva} \varepsilon_w = \frac{\dot{Q}}{h_{fg}} \tag{1}$$

3.1 Energy conservation and temperature profile in the evaporator

The energy conservation equations of the single wick structure are given by [4, 25]

$$\frac{\dot{m} C_{pl}}{2\pi r L_{eva}} \frac{\partial T_w}{\partial r} = k_{eff} \frac{1}{r} \frac{\partial}{\partial r} \left(\mathrm{r} \frac{\partial \mathrm{T_w}}{\partial r} \right) \tag{2}$$

The effective thermal conductivity of liquid-saturated wick in a cylindrical geometry is [3, 4]

$$k_{eff} = \frac{k_l[(k_l + k_w) - (1 - \varepsilon_w)(k_l - k_w)]}{[(k_l + k_w) - (1 - \varepsilon_w)(k_l - k_w)]} \tag{3}$$

in which, the porosity of screen wick is expressed as [3, 4]

$$\varepsilon_w = 1 - \frac{1.05\pi n_w D_w}{4} \tag{4}$$

Define the variable α as

$$\alpha = \frac{\dot{m} C_{pl}}{2\pi r L_{eva} k_{eff}} \tag{5}$$

Then, rewrite Eq. (2) as

$$\frac{\partial^2 T_w}{\partial r^2} + \frac{1}{r}(1 - \alpha)\frac{\partial \mathrm{T_w}}{\partial r} = 0 \tag{6}$$

The boundary conditions are

$$\begin{cases} T(r)|_{r=r_{w,0}} = T_{w,0} \\ T(r)|_{r=r_{w,i}} = T_{w,i} \end{cases} \tag{7}$$

By solving the second-order ordinary differential Eq. (6) with twice integrals, temperature distribution in the wick is

$$T(r) = \left[\frac{T_{w,0} - T_{w,i}}{(r_{w,0} - r_{w,i})^{\alpha} - 1}\right] \times \left(\frac{r}{r_{w,i}}\right)^{\alpha} - \frac{T_{w,0} - T_{w,i}(r_{w,0} - r_{w,i})^{\alpha}}{(r_{w,0} - r_{w,i})^{\alpha} - 1} \tag{8}$$

For a single saturated wick layer, the radial thermal conductance is then

$$G_w = \frac{k_{eff} A_{w,i}}{T_{w,0} - T_{w,i}} \times \left.\frac{\partial T_w}{\partial r}\right|_{r=r_{w,i}} = \frac{\dot{m} C_{pl}}{(r_{w,0} - r_{w,i})^{\alpha} - 1} \tag{9}$$

In this case, the temperature distributions are

$$\begin{cases} r_3 \le r \le r_2, T(r) = \left[\frac{T_{w,2} - T_{w,3}}{(r_2/r_3)^{\alpha_2} - 1}\right] \times \left(\frac{r}{r_3}\right)^{\alpha_2} - \frac{T_{w,2} - T_{w,3}(r_2/r_3)^{\alpha_2}}{(r_2/r_3)^{\alpha_2} - 1} \\ r_2 \le r \le r_1, T(r) = \left[\frac{T_{w,1} - T_{w,2}}{(r_1/r_2)^{\alpha_1} - 1}\right] \times \left(\frac{r}{r_2}\right)^{\alpha_1} - \frac{T_{w,1} - T_{w,2}(r_1/r_2)^{\alpha_1}}{(r_1/r_2)^{\alpha_1} - 1} \\ r_1 \le r \le r_0, T(r) = T_{w,1} + \frac{(T_{eva,wall} - T_{w,1})}{\ln\left(\frac{r_0}{r_1}\right)} \ln\left(\frac{r}{r_1}\right) \end{cases} \tag{10}$$

Thermal conductance of the inner and outer wick layers are respectively

$$\begin{cases} G_{w,0} = \frac{k_{eff,0} A_{w,2}}{T_{w,1} - T_{w,2}} \times \left.\frac{\partial T_w}{\partial r}\right|_{r=r_{w,2}} = \frac{\dot{m} C_{pl}}{(r_1/r_2)^{\alpha_1} - 1} \\ G_{w,i} = \frac{k_{eff,i} A_{w,3}}{T_{w,2} - T_{w,3}} \times \left.\frac{\partial T_w}{\partial r}\right|_{r=r_{w,3}} = \frac{\dot{m} C_{pl}}{(r_2/r_3)^{\alpha_2} - 1} \end{cases} \tag{11}$$

and

$$\begin{cases} \alpha_1 = \frac{\dot{m} C_{pl}}{2\pi k_{eff,0} L_{eva}} \\ \alpha_2 = \frac{\dot{m} C_{pl}}{2\pi k_{eff,i} L_{eva}} \end{cases} \tag{12}$$

$$\begin{cases} k_{eff,0} = \frac{k_l[(k_l + k_{w,0}) - (1 - \varepsilon_{w,0})(k_l - k_{w,0})]}{[(k_l + k_{w,0}) - (1 - \varepsilon_{w,0})(k_l - k_{w,0})]} \\ k_{eff,i} = \frac{k_l[(k_l + k_{w,i}) - (1 - \varepsilon_{w,i})(k_l - k_{w,i})]}{[(k_l + k_{w,i}) - (1 - \varepsilon_{w,i})(k_l - k_{w,i})]} \end{cases} \tag{13}$$

As a result, the overall thermal conductance of the composite wick structure can be given by

$$G_w = \frac{k_{eff,i} A_{w,3}}{T_{w,1} - T_{w,3}} \times \left.\frac{\partial T_w}{\partial r}\right|_{r=r_{w,3}} = \frac{T_{w,2} - T_{w,3}}{T_{w,1} - T_{w,3}} \times \frac{k_{eff,i} A_{w,3}}{T_{w,2} - T_{w,3}} \times \left.\frac{\partial T_w}{\partial r}\right|_{r=r_{w,3}} = \frac{T_{w,2} - T_{w,3}}{T_{w,1} - T_{w,3}} \times G_{w,i} \tag{14}$$

According to energy conservation, heat flux at the internal surface of the outer wick layer should be equal to the heat flux at the external surface of the inner wick layer

$$k_{eff,0}\frac{\partial T_w}{\partial r}\bigg|_{r=r_{w,2+}} = k_{eff,i}\frac{\partial T_w}{\partial r}\bigg|_{r=r_{w,2-}} \tag{15}$$

Putting Eq. (10) into the above expression results in

$$\frac{T_{w,1}-T_{w,2}}{(r_1/r_2)^{\alpha_1}-1} = \left[\frac{T_{w,2}-T_{w,3}}{(r_2/r_3)^{\alpha_2}-1}\right] \times \left(\frac{r_2}{r_3}\right)^{\alpha_2} \tag{16}$$

Put Eq. (11) into Eq. (16) and it becomes

$$G_{w,0}(T_{w,1}-T_{w,2}) = G_{w,i}(T_{w,2}-T_{w,3}) + \dot{m}C_{pl}(T_{w,2}-T_{w,3}) \tag{17}$$

Rewrite Eq. (17) as

$$T_{w,2} = \frac{G_{w,0}T_{w,1} + \left(G_{w,i}+\dot{m}C_{pl}\right)T_{w,3}}{G_{w,0}+G_{w,i}+\dot{m}C_{pl}} \tag{18}$$

Put Eq. (18) into Eq. (14) for obtaining the expression of the composite wick structure,

$$G_w = \frac{G_{w,0}G_{w,i}}{G_{w,0}+G_{w,i}+\dot{m}C_{pl}} \tag{19}$$

The thermal resistance in this region is therefore

$$R_{eva} = \frac{\ln\left(\frac{r_0}{r_1}\right)}{2\pi L_{eva}k_{eva,wall}} + \frac{1}{G_w} \tag{20}$$

The interface temperature conditions can be assumed local thermal equilibrium

$$T_{w,3} = T_v = T_{int} \tag{21}$$

3.2 Flow characteristic

In a heat pipe, the maximum capillary pumping head ($\Delta P_{c,\max}$) must be greater than or at least equal to the total pressure drops (ΔP) along the heat pipe. The total pressure drops (ΔP) should be the sum of pressure drops in all the heat pipe components, that is, wick structure, evaporator, three-way separator, vapour line, condenser and liquid line.

$$\Delta P_{c,\max} + \Delta P_g \geq \Delta P \tag{22}$$

$$\Delta P = \Delta P_w + \Delta P_{eva} + \Delta P_{tw} + \Delta P_{vl} + \Delta P_{cond} + \Delta P_{ll} \tag{23}$$

4. Design and fabrication of the proposed GALHP

Based on the results derived from the theoretical and computer simulation studies [26], the proposed GALHP was designed, fabricated and presented in **Figure 3** respectively. For the evaporator, the length remained 550 mm and diameter fixed to 22 mm. Within the inner surfaces of the evaporator, the compound screen mesh wick structure was applied with the size of 160 × 60 mm.

For the condenser, they were all fixed with a steel cooling jacket of the same size, with a length of 150 mm and a diameter of 105 mm. The detailed design parameters are illustrated in **Table 1**.

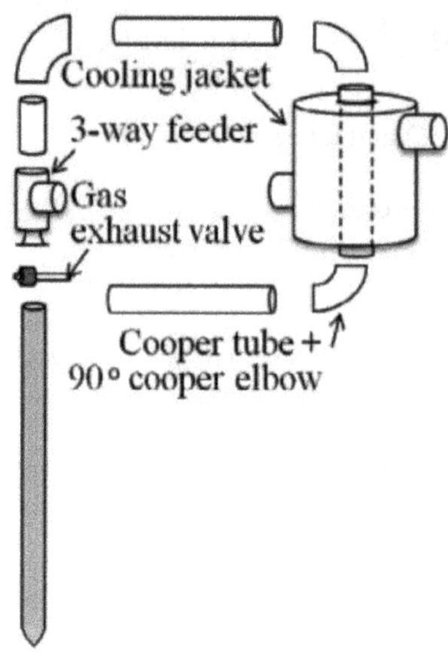

Figure 3.
Fabrication schematics of the proposed GALHP.

Parameters	Nomenclature	Value
External diameter of evaporator (mm)	$D_{hp,o}$	22
Internal diameter of evaporator (mm)	$D_{hp,in}$	19.6
T1: internal diameter of vapour channel in three-way fitting (mm)	$D_{tw,in}$	14
Operating pressure in heat pipe (Pa)	P_{hp}	1.3×10^{-4}
Evaporator length (mm)	$L_{hp,ev}$	550
Liquid filling volume (ml)	V_{fl}	85
Transportation line's outer diameter (mm)	$D_{ll,o}/D_{vl,o}$	22
Transportation line's inner diameter (mm)	$D_{ll,in}/D_{vl,in}$	19.6
Lengths of vapour/liquid /line (mm)	$L_{vl,T1}/L_{ll,T1}$	595/445
	$L_{vl,T2}/L_{ll,T2}$	595/1145
Mesh screen wire diameter (layer I) (mm)	$D_{owi,ms}$	7.175×10^{-2}
Mesh screen layer thickness (layer I) (mm)	$d_{owi,ms}$	3.75×10^{-1}
Mesh number (layer I) (/m)	$N_{owi,ms}$	6299
Mesh screen wire diameter (layer II) (mm)	$D_{iwi,ms}$	12.23×10^{-2}
Mesh screen layer thickness (layer II) (mm)	$d_{iwi,ms}$	3.75×10^{-1}
Mesh number (layer II) (/m)	$N_{iwi,ms}$	2362
Mesh screen conductivity (W/m °C)	k_{ms}	394

Table 1.
Design parameters of the proposed GALHP.

5. Experimental set-up and procedure

5.1 Experimental set-up and instrumentation

Figure 4 shows the test rig of the proposed GALHP. In the rig, an electrical heating tap with the percentage controller, which acts as the heat source, was evenly attached to the external surfaces of the evaporators. The condenser is covered by a steel cooling jacket that allows cooling water to pass through, removing heat from the condenser. A magnetic regeneration water pump was installed in the cooling water loop to power the cooling water cross. A clamp-supported retort stand was used to adjust the inclination angle of the piping installation. The foamy polyurethane was attached to the pipes to provide a satisfactory insulation. During operation, in order to keep a relatively constant condensation temperature, the water tap would remain open to enable adequate amount of cold water to be fed into the loop. When the water tank was fully charged, the drainage valve would be turned open to allow the extra amount of water to be discharged.

A list of the piping elements and test instruments are provided in **Table 2**. A number of T-type thermocouples were attached to the external surface of heat pipe walls, and installed in the inlet/outlet and inside of cooling jacket and water tank: there were totally four thermocouples (No. 1–4) equidistantly attached along each heat pipe evaporator wall from top to bottom, which were used to measure the temperature distribution along the evaporator wall and their corresponding average temperature at the evaporation sections; another four thermocouples were respectively placed in the mid of heat pipe condenser wall (No. 7), the inlet/outlet

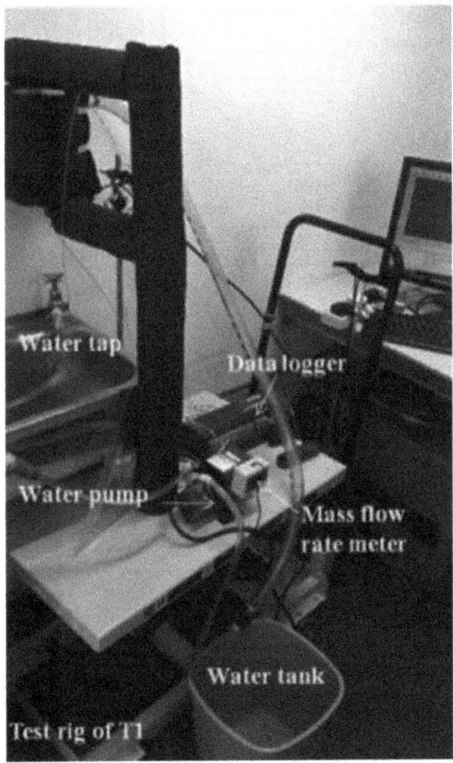

Figure 4.
On-site testing rig.

No.	Name	Model no.	Description
1	PVC-U ball valve	RS: 282-5148	Compression fitting size: 20mm
2	PVC-U hose connector	RS: 212-3638	1/2in BSPT MX20 mm
3	George fischer 90° PVC-U elbow	RS: 279-0575	25 × 25 mm, L. 33 mm
4	Armorvin HNA hose	RS: 339-9921	Clear 5 m × 20 mm ID
5	Magnetically coupled regenerative pump	Totton: HPR6/8	Max capacity: 5.5 l/min Max head: 7.4 m
6	Heating tapes with percentage controller	Omega: HTWC102-004	Length: 1220 mm 14.4–288 w; 240 V; ±0.5 W
7	Black nitrile rubber pipe insulation	RS: 486-053	Ø 22 × 25 mm
			k = 0.035 W/m °C at 0°C
			0.037 W/m °C at 20°C
			0.040 W/m °C at 40°C Min/max temperature sensed
8	T-type thermocouple	RS: 621–2164	Min/max temperature sensed: ±0.1°C, 200–350°C Probe diameter: 0.3 mm
9	1/2" LCD Water fluid flow sensor meter digital display rate turbine flow meter		Flow rate range: 1.5–25.0 l/m, fitting for 1/200, BSP: ±0.1 l/m, water temperature range: 0–80°C
10	Data logger and data recording equipment	TD500 series 3	10 channels DataTaker; ±0.16% 5-s interval recording

Table 2.
List of piping connectors and experimental instruments.

Test mode	Applied heat load (W)	Inclination angle (°)	Cooling-fluid temperature (°C)	Flow rates of cooling fluid (l/min)
Standard	100.8	90	10 ± 0.5	1 ± 0.2
1	14.4	90	10 ± 0.5	1 ± 0.2
	43.2			
	100.8			
	129.6			
2	100.8	90	10 ± 0.5	1 ± 0.2
		60		
		30		
3	100.8	90	10 ± 0.5	1 ± 0.2
			15 ± 0.5	
			20 ± 0.5	
4	100.8	90	10 ± 0.5	1 ± 0.2
				2 ± 0.2
				3 ± 0.2

Table 3.
List of operational modes (parameters) for simulation of the three heat pipes.

Q (W)	u (°)	T_{cf} (°C)	m_{cf} (l/m)	T_{eva} (exp) (°C)	T_{eva} (sim)	CR (−)	E (%)	U (%)	R (exp) (°C/W)	R (sim)	CR (−)	E (%)	U (%)	Start-up (s)
14.4	90	10	1	31.31	34.12	0.8346	6.61	8.05	0.32	0.32	0.9533	18.18	10.63	840
43.2				40.04	36.41			7.34	0.27	0.21			8.26	780
129.6				42.11	43.28			8.00	0.08	0.09			10.26	280
100.8				41.14	40.99			6.80	0.11	0.10			9.46	410
	60			46.72	45.65	0.9825	1.30	5.65	0.13	0.15	0.9995	11.88	6.19	960
	30			48.90	49.43			5.97	0.16	0.20			11.47	1210
	90	15		56.53	48.65	0.8827	8.14	9.98	0.15	0.18	0.9855	13.76	12.36	370
		20		56.92	55.99			7.94	0.14	0.17			9.42	320
		10	2	40.96	38.65	0.9413	6.26	8.89	0.11	0.09	0.9556	13.71	10.65	380
			3	39.95	35.99			10.81	0.10	0.08			11.76	360

Table 4.
Comparison of testing and simulation results of the three heat pipes under different modes.

(No. 5/6) of cooling jacket and the mid of water tank (No. 8) to measure their related average temperature; there were still four more thermal couples applied to measure the temperatures at the vapour (No. 9, 11) and liquid (No. 10, 12) transportation lines; all these thermocouples were further connected to a data logger to record the temperature signals at each testing interval. A control box is provided to adjust the power output of the heating belt, which is considered a thermal load.

5.2 Experimental process

A series of laboratory steady-state tests were carried out and the results of the tests were used to evaluate the thermal performance of the proposed GALHP. The testing conditions are displayed in **Table 3**. During all the sets of tests, the surrounding air temperature and speed were maintained at 20 ± 2°C and 0.01 m/s, respectively. Under initial test conditions, one parameter is changed and the other parameter remains fixed, enabling the development of a correlation between the heat pipe's heat output and associated operating parameters. Once the steady-state conditions have been achieved, the test period is successive 10 h period. The measurement data will be recorded at 5-s interval and logged into the computer system using the DT500 data logger to enable the follow-up analyses to be undertaken.

6. Computer model validation using the experimental results

Table 4 provides the comparison between the testing and the simulation results under all selected testing conditions. The mean correlation coefficient (CR) was found no less than 0.8346 and the root mean square percentage deviation (E) was below 18.18%. This indicated that the developed simulation model could predict the thermal performance at a reasonable accuracy. The differences resolved above may be caused by theoretical and/or inaccurate measurements. From the theoretical side, some simplified assumptions and empirical equations were involved; from the experimental side, a few of the uncertainties addressed above may be the potential reasons for the deviation. Based on these considerations, the errors may be attributed to the theoretical inaccuracies and it would be better for the simulation model to be refined to further improve its accuracy in making predictions based on the experimental results.

7. Conclusion

This chapter reported the study of a novel liquid-vapour separator-incorporated gravity-assisted loop heat pipe (GALHP), which was designed, constructed, and tested. A parallel comparison between simulation and experimental results was made.

Under the specified operational conditions, the start-up timing of the proposed GALHP was 410 s. The overall thermal resistance was 0.11°C/W, indicating that it has small heat transfer resistance owing to its unique structure that led to the even liquid film distribution and thus reduced flow resistance. The actual effective thermal conductivity was 29,968 W/°C m, indicating that it achieved significant improvement in terms of heat transfer. All of these data provide evidence that the proposed GALHP is a super-performance heat transfer device that can be widely used in gravity-assisted heat transfer operations to achieve significant thermal management in a variety of practical applications.

Author details

Xudong Zhao[1*], Chuangbin Weng[2], Xingxing Zhang[3], Zhangyuan Wang[2] and Xinru Wang[4]

1 School of Engineering, University of Hull, UK

2 School of Civil and Transportation Engineering, Guangdong University of Technology, Guangzhou, China

3 Department of Energy, Forest and Built Environments, Dalarna University, Falun, Sweden

4 Department of Architecture and Built Environment, University of Nottingham, Ningbo, China

*Address all correspondence to: xudong.zhao@hull.ac.uk

References

[1] Wang Z, Yang W. A review on loop heat pipe for use in solar water heating. Energy and Buildings. 2014;**79**:143-154

[2] Zhang H, Zhuang J. Research, development and industrial application of heat pipe technology in China. Applied Thermal Engineering. 2003;**23**: 1067-1083

[3] Amir F. Heat Pipe Science and Technology. 1st ed. UK: Taylor & Francis Group; 1995

[4] David R, Peter K. Heat Pipes: Theory, Design and Applications. 5th ed. UK: Elsevier; 2006

[5] Maidanik YF. Loop heat pipes. Applied Thermal Engineering. 2005;**25**: 635-657

[6] Dunn PD, Reay DA. The heat pipes. Physics in Technology. 1973;**4**:187-201

[7] Reay D, Kew P. Heat Pipe. 5th ed. London, UK: Elsevier; 2006

[8] Xu X, Wang S, Wang J, Xiao F. Active pipe-embedded structures in buildings for utilizing low-grade energy sources: A review. Energy and Buildings. 2010;**42**:1567-1581

[9] Maidanik YF, Vershinin S, Kholodov V, Dolgirev J. Heat Transfer Apparatus, US patent 4515209; 1985

[10] Peterson GP. An Introduction to Heat Pipes: Modelling, Testing, and Applications. New York: Wiley-Interscience Press; 1994

[11] Ratios of Specific Heat. 2009. Available from: http://www.engineeringtoolbox.com/specific-heat-ratio-d608.html [Accessed: November 27, 2009]

[12] Zuo ZJ, Faghri A. A network thermodynamic analysis of the heat pipe. International Journal of Heat and Mass Transfer. 1998;**41**:1473-1484

[13] Kaya T, Hoang TT. Mathematical Modelling of Loop Heat Pipes; American Institute of Aeronautics and Astronautics (AIAA), Paper No. AIAA 99-04771999. pp. 1-10

[14] Bai L, Lin G, Wen D. Modelling and analysis of start-up of a loop heat pipe. Applied Thermal Engineering. 2000;**30**: 2778-2787

[15] Pauken M, Rodriguez JI. Performance Characterisation and Model Verification of A Loop Heat Pipe; Society of Automotive Engineers (SAE), Paper No. 2000-01-0108; 2000

[16] Hoang TT, Cheung KH, Baldauff RW. Loop Heat Pipe Testing and Analytical Model Verification at the US Naval Research Laboratory, SAE International Paper No. 04ICES-288; 2004

[17] Riehl RR. Comparing the Behaviour of a Loop Heat Pipe with Different Elevations of the Capillary Evaporator, SAE International Paper No. 2004-01-2510; 2004

[18] Zan KJ, Zan CJ, Chen YM, Wu SJ. Analysis of the parameters of the sintered loop heat pipe. Heat Transfer - Asian Research. 2004;**33**:515-526

[19] Riehl RR, Dutra T. Development of an experimental loop heat pipe for application in future space missions. Applied Thermal Engineering. 2005;**25**: 101-112

[20] Vlassov VV, Riehl RR. Modelling of a Loop Heat Pipe for Ground And Space Conditions, SAE International Paper No. 2005-01-2935; 2005

[21] Launay S, Platel V, Dutour S, Joly J-L. Transient modelling of loop

heat pipes for the oscillating behaviour study. Journal of Thermophysics and Heat Transfer. 2007;**21**:487-495

[22] Zhang X et al. Characterization of a solar photovoltaic/loop-heat-pipe heat pump water heating system. Applied Energy. 2013;**102**:1229-1245

[23] Zhang X et al. Dynamic performance of a novel solar photovoltaic/loop-heat- pipe heat pump system. Applied Energy. 2014;**114**: 335-352

[24] He W et al. Theoretical investigation of the thermal performance of a novel solar loop-heat-pipe facade-based heat pump water heating system. Energy and Buildings. 2014;**77**:180-191

[25] Rohsenow W, Hartnet J, Cho Y. Handbook of heat transfer. 3rd ed. New York, USA: McGraw-Hill; 1998

[26] Zhang X et al. Comparative study of a novel liquid-vapour separator incorporated gravitational loop heat pipe against the conventional gravitational straight and loop heat pipes—Part I: Conceptual development and theoretical analyses. Energy Conversion and Management. 2015;**90**: 409-426

www.ingramcontent.com/pod-product-compliance
Lightning Source LLC
Chambersburg PA
CBHW070242230326
41458CB00100B/5909
* 9 7 8 1 8 3 9 6 2 6 4 3 2 *